THE ASSOCIATION FOR
SCIENCE EDUCATION

Alternatives
for
Science
Education

A CONSULTATIVE DOCUMENT

1979

The Association for Science Education
College Lane, Hatfield, Herts AL10 9AA

©ASE 1979

ISBN 0 902 786 49 0

Made and printed in Great Britain by
The Garden City Press Limited
Letchworth, Hertfordshire SG6 1JS

CONTENTS

3

CONTENTS

PREFACE

This consultative document has been prepared by a small Working Party established under the auspices of the Education (Research) Committee of the Association for Science Education. The Working Party represents a range of views within the full spectrum of interests in the Association. In spite of their diversity of employment and opinion they share a common interest, namely the establishment of a forward-looking science education policy through the 1980s.

The Education Committee wishes there to be a broadly based debate within the membership. This document will be sent to every member of the Association and it is hoped that a pattern of individual, local, regional and national responses will lead in a structured way to the development of a policy statement to which the great majority of the membership could subscribe.

The Committee wish to record their debt of gratitude to the members of the Working Party for the time, effort and devotion spent in their deliberations.

E. C. Ellington, The Piggott School, Wargrave-on-Thames
J. G. Ellis, Paddington College, London
K. Hinton, Clissold Park School, London
C. H. Johnson, Whitchurch High School, Cardiff
D. G. Kincaid, Buckinghamshire County Council
J. J. Thompson (Chairman), Department of Educational Studies, University of Oxford
R. W. West, School of Education, University of Sussex

Finally, thanks are due to Mrs Gwyneth White and Mrs Felicity Hanley for their help and patience in providing secretarial support.

M. J. SAVORY
(Chairman, Education (Research) Committee)

INTRODUCTION

The aim of this document is to present alternative courses of action that may be followed by those concerned with the place of science in education. It has been prepared by a working party comprised of members of the Association for Science Education who, between them, have experience of most sectors of education, with considerable help from other members whose views and suggestions are gratefully acknowledged. We hope that our contribution will be of value both to educational administrators who, in implementing plans and providing resources, need to consider the place of science in the curriculum, and also to science teachers whose willing and informed participation is essential.

Our report contains three main sections and a summary of our conclusions. The first section is a *historical review* that attempts to identify those major strands in the development of British education which we believe account for the state of education in the late 1970s. Such a review, if only because of its brevity, must be selective and subjective, and the reader may not entirely agree with our analysis. We hope that the majority of what we say will meet with approval, and that nothing will meet with universal condemnation. Secondly, we make a *survey of the present*, and the need to reduce our survey to manageable proportions has compelled us to omit much of the finer detail. By painting the scene in broad strokes only, we hope that we have not given a distorted view, but only a simplified one which is acceptable. Thirdly, we give some of the *options for the future*, based on several assumptions about our economic and political future. We are not arguing for any one of our options, but only presenting the choices as we see them. We hope that discussion of these options will follow, and that policies will then be formulated to determine the development of science education until the turn of the century, and maybe beyond. That there is a need for policies cannot be doubted; the Association wishes to play its part in formulating these policies.

The Association for Science Education is the largest subject teaching association in the United Kingdom, with over 16 000 members concerned with all aspects and all levels of science education. Administrators, advisers, technicians and teachers in all types of school and college are members, the largest category being teachers in primary and secondary schools. In addition to providing a forum for discussion of good "classroom practice", the Association maintains links with industry, such bodies as the Royal Society and the British Association, the professional Institutes of Biology, Chemistry, Physics, Engineering and Applied Sciences, the teacher training institutions and the universities and polytechnics, HM Inspectorate and LEA advisers (many of whom are active members of the Association), the Schools Council, educational publishers and apparatus manufacturers and with other subject associations, both in the UK and internationally. From time to time the Association publishes policy documents, the significance of which may be judged from the impact of the 1957 documents [1] which led to the establishment in 1962 of the Nuffield Foundation

7

Science Teaching Project and the consequent curriculum developments of the 1960s. The social and educational climates are now very different from those of 20 years ago; with this report we offer a new basis for discussion, negotiation and action. If we make only a fraction of the impact of earlier reports, we shall be well satisfied.

STRUCTURE OF THE REPORT

Part One of this report is concerned with an analysis of the various **contexts** from which various **proposals** will be derived in Parts Two and Three. Part One represents the foundations for our proposals and has been written with both theory and practice in mind. Thus we are concerned with **three** distinct but interrelated contexts, namely:

(a) changing views on the nature of educational processes and provision (educational context);

(b) an analysis of changes in the society that supports educational institutions and practice (social context); and

(c) an analysis of the range of scientific, psychological and sociological theories that influence educational practice and social interpretations (theoretical context).

We then present in Part Two of the report a review, and critique, of current curriculum provision together with an analysis of the resource implications of a future model of the science curriculum not too different from those at present in existence.

In Part Three of the report we anticipate possible developments in the structure of the science curriculum, and identify three curriculum models which represent short-, medium- and long-term goals respectively. Finally, the implications of such proposals for the future are set out.

Part One   Science Education in Context

In order to understand the present position of science education in our schools we feel it is necessary to consider the educational, social and theoretical background to current provision.

EDUCATIONAL CONTEXT

Introduction

The last major policy statement of the Association for Science Education was prepared when educational provision was still dominated by the tripartite system of secondary education. Selection and rigid streaming were general practice, and between 70 and 80 per cent of young people left school at 15. Opportunities for further, higher or adult education were very limited. Primary education in most parts of the country was dominated and severely constrained by the demands of the 11+ examinations and the need to provide a common experience and base for two, if not three, quite distinct patterns of secondary education. The primary school curriculum in the 1950s contained virtually no provision for the study of science. Secondary education for all had, following the 1944 Education Act, become a reality, but the reality was differently interpreted in the grammar, technical and modern schools maintained by most local authorities. Apart from the then recently introduced General Certificate of Education O- and A-level examinations designed for the grammar schools, no coherent examination system existed for the majority of secondary school children. Transfer between schools at 13+ and 15+ existed as a difficult and often stressful possibility for late developers and pupils wrongly assessed at 11+. The preparatory, public and direct grant schools provided a powerful and heavily utilized alternative system to that of the maintained sector, a system more closely identifiable with grammar school traditions than those of the technical and modern schools established post-1944.

Further education provision, limited by the lack of adequate day-release opportunities, existed in almost total isolation from the secondary schools and had few effective links with higher education. Higher education in the 1950s catered for a very small minority of able boys and girls. In 1957 there were 24 universities in the UK providing advanced educational opportunities for only 3.9 per cent of 18-year-olds [2].

The science curriculum in the 1950s reflected the attitudes and values of the public and grammar schools and their close and exclusive links with the universities. Science education was subject-focused and essentially conceptualized in terms of O- and A-level courses in biology, chemistry and physics with the option of physics-with-chemistry for some pupils. Many boys' grammar schools failed to provide adequate opportunities for the study of biology and the amount of physical sciences taught in girls' schools was a cause of concern at the time. General science, following a chequered history of fits and starts, had in no sense established itself in the grammar school curriculum. A comparison of O- and A-level passes in the GCE examinations for 1952 and 1975, Table 1, shows the small number of pupils successfully completing courses in 1952 and the sex differences in attainment between different subjects. It also shows the number of pupils passing O- and A-level examinations in 1975 compared with the numbers doing so in 1952 [3, 4].

Table 1. Number of passes in GCE O- and A-level science examinations in 1952 and 1975

Subject		O level 1952	O level 1975	A level 1952	A level 1975
Physics	(B)	12 720	61 406	8 925	23 954
	(G)	1 715	17 121	1 258	5 342
	(T)	14 435	78 527	10 183	29 296
Chemistry	(B)	11 838	45 087	7 743	16 891
	(G)	3 006	21 133	1 372	6 937
	(T)	14 844	66 220	9 115	23 828
Biology	(B)	5 283	45 972	4 390[c]	11 616
	(G)	15 002	68 471	2 714	11 407
	(T)	20 285	114 443	7 104	23 023
Other	(B)	3 135[a]	66 489[b]		
	(G)	1 350	10 103	Not given	
	(T)	4 485	76 592		

Notes
[a] Physics-with-Chemistry.
[b] Other science or technical subjects.
[c] Shown separately as Biology, Botany and Zoology in 1952 tables but consolidated here.

Compared with today, science education in the 1950s was a small-scale business, organised by a small number of highly dedicated teachers of chemistry, biology and physics whose horizons were limited by the tight demands of the subject disciplines and the particular needs of university science departments. Science education had no significant role in general education, and large numbers of pupils left school without an adequate background of science studies.

It would, however, be a gross over-simplification to leave an analysis of the 1950s as cut and dried as this. The secondary modern schools, established in the euphoria of universal secondary education arising from the 1944 Education Act, had, by the mid-fifties, started to establish their own traditions of science teaching. Teacher training colleges were sufficiently isolated from the academic traditions of the universities to have developed their own solution to the social and educational problems of the schools they served. In general, college science departments identified their task as one of providing science teachers for secondary modern schools: teachers who would attempt to teach ordinary children a form of science studies that was both interesting and relevant. In so far as a philosophy for science education existed, it could be said to reflect an overt concern with the child as the central focus for educational planning, a preoccupation with the science of everyday things and everyday life [5, 6]. and an active concern with theories of child psychology and learning [7].

This emerging and, as it happened, short-lived tradition of secondary modern science teaching, wrongly equated in recent years with the issue of general or

12

integrated science, contrasted starkly with the established traditions of the grammar and public schools mentioned above. In the grammar schools, the main concerns of both students and teachers were related to the processes whereby the young could be socialized into the thought and behaviour patterns of the "elders"; a highly professional activity governed by a tightly hierarchical relationship between the pupil, the teacher and the university admissions tutor. It was, furthermore, an activity governed by very strict academic rules as to what constituted science and scientific behaviour. In the modern schools the child was the central focus; an individual to whom science could be in part explained and with whom in part used; a highly functional approach not in any real sense constrained by the structure or history of the subject.

It is vitally important to an understanding of the contextual situation in the fifties, the time of the last major Policy Statement, to realize that both the grammar school and secondary modern school traditions made their own, if very different, demands on both the science teacher and the science educator. The former tradition placed a high premium on academic scholarship involving a deep and systematic concern for the *content* of science; a concern with the latest developments in the subject; and a search for the ways and means whereby new knowledge could be incorporated into teaching schemes. Leaving status issues aside, the alternative tradition was much more overtly concerned with pedagogy; the study of ways and means whereby science, in whatever form, could be effectively mediated between teacher and taught. In the 1950s both traditions lacked a clear analysis of *process*; the analysis of the nature of science and how it is created. In short, both groups were faced with complementary problems—what to teach, and how best to teach it.

At this time the level of support for science teachers provided by the Science Masters' Association and the Association of Women Science Teachers was very limited and almost totally concerned with science teaching in the grammar and public schools. The combined membership of the two associations was, in 1957, approximately 6 620 with only a token representation of teachers working in primary schools, modern schools and further education. In addition, few LEAs provided effective specialist advisory support for their science teachers and such provision as existed was dominated by the concept of the "Inspector".

The curriculum development movement of the 1960s, typified by the work of the Nuffield Science Teaching Project, was an overt attempt to resolve both the content and pedagogical aspects of the science curriculum. For political, economic and defence reasons attention had been focused, on both sides of the Atlantic, on the low level of investment in, and recruitment to, science and technology. Industry, in its various forms, came to the rescue of the schools by funding:

(a) through the aegis of the "Industrial Fund", the building and equipment of public school science laboratories, sadly neglected following generations of commitment to the classics; and

13

(b) the Nuffield Science Teaching Project, which has at the secondary level concentrated its main energy on the education of the future scientist.

The secondary modern school tradition, which lacked status, support and clear articulation, received very short shrift and to date few major projects have been funded which seek to explore and further develop a programme of science education based on these alternative principles, and the emerging concept of "science for all". Such shifts in emphasis that have occurred have been largely limited to a revision of the content of school science education and marginal changes in the method of its presentation. We shall return to this issue later in this report.

Organizational change

Concurrent with these changes in the content of science education programmes there has been a major revolution in the organizational *context* within which science is taught. During the late fifties and early sixties some local education authorities began to develop alternative systems of secondary education, and by the late sixties comprehensive secondary education had become a central component of government policy.

The effects of Circular 10/65 and the gradual removal of 11+ selection freed primary education from many of its earlier constraints and has led to a wide range of curricular and organizational reforms. Following the *Plowden Report* [8] LEAs, on a combination of educational and economic considerations, have developed a variety of first and middle school systems which have raised major questions about the nature of the curriculum in the middle years (8–13). It can, for example, no longer be assumed that pupils aged 11 will, or should, switch from a teacher-centred integrated curriculum to a subject-centred differentiated curriculum. Furthermore, the removal of the constraint of the 11+, with its inevitable emphasis on a limited range of basic skills, has led in most primary schools to a broadening of the curriculum particularly in the fields of creative work, language and basic science studies. A key policy issue for the ASE may well be the problem of effectively defining the objectives, content and approach of science studies in the middle years [9].

At secondary level the movement towards comprehensive education has, again due to a combination of educational and economic considerations, led to a diversity of provision, not all of which can be comfortably accommodated under the general formula of "good educational practice". Provision ranges from all-through 11–18 schools, to various combinations of junior and high schools, and 11–16 secondary schools feeding sixth-form colleges. In addition LEAs have adopted differing policies regarding both pupil allocation and the relationship between schools and the community. The definition of catchment areas and the various interpretations of the concept of parental choice have led to the creation of both large and small comprehensive schools; neighbourhood schools with either very wide or very narrow levels of social and ethnic mixing; inner urban schools located with small catchment areas contrasting with others serving vast rural

14

catchment areas. Similarly, while many secondary comprehensive schools exist in isolation from the immediate concerns and needs of the local community, a growing number are becoming closely integrated as centres for educational, cultural and recreational activities for people of all ages living and working in the locality. The village colleges developed by William Morris in Cambridgeshire now have their urban counterparts.

In our general consideration of the results of comprehensive reorganization we should note the wide variety of internal organization that exists between schools. In some, tripartite assumptions regarding abilities, aptitudes and interests have not been questioned and pupils, teachers and the timetable reflect sharp differentiation between streams, bands, courses and options. In others, de-streaming, mixed ability grouping, faculty organization and complex patterns of core and optional studies suggest that alternative educational rationales have been established. This diversity of provision and practice reflects a variety of political and value positions as well as definitions of science and education. It is obviously dangerous to generalize about the problems of comprehensive secondary education from the basis of practice in any one school.

One of the most significant developments to occur concurrently with the movement towards comprehensive reorganization has been the creation of the CSE examination system following the *Beloe Report* in 1960 [10]. This flexible system, controlled by the teaching profession itself, has also contributed to the variety of provision in secondary schools in its alternative modes of assessment. While the ASE in recent years has increased its involvement in the problems of the comprehensive school [11], we are conscious that a research and development programme based on a clearly articulated policy and ordering of priorities must be developed if the Association is to increase its influence in the field of secondary education. This links with the evolution of the Schools Council from the Secondary Schools Examinations Council in 1964. The Schools Council has become a major agency for curriculum and examination reform and has taken its share of responsibility for much of the curriculum development in science education in the 1960s and 1970s.

The independent sector
The preparatory and public school system, which has continued to expand and improve its science programmes, remains a powerful alternative to the maintained sector. Science education in the non-maintained sector, which benefits from high levels of internal support and a willingness to innovate, still maintains its special relationship with higher education and industry. The private sector, however, in spite of its high status and privilege, may not be able in the long term to resist the social pressures and theoretical issues identified in later sections of this part of the report.

Further education
Post-1960s developments in further education have shown an increase in the variety of provisions and an expansion in the number of institutions, courses and

places. Further education is moving into a closer and more formalized relationship with both secondary and higher education, a process that may well be hastened by the general reorganization of educational provision for the 16–19-year-olds.

In further education colleges the provision of full-time courses leading to GCE at both O and A level and to Ordinary National Diplomas has been recognized as an increasingly viable alternative to the school sixth form as a route into higher education. Those who leave school at 16+ to enter employment have been able to use part-time GCE or Ordinary National Certificate courses to gain subsequent access to full-time higher education at universities or polytechnics. Alternatively, they have undertaken part-time higher education courses leading to degrees or Higher National Certificates with subsequent membership of the professional institutions. It is important to recognize that a considerable proportion of the further education provision has been at craft or operative level, providing good opportunities for students who failed in the more élitist and academic atmosphere which they had encountered in school. The more adult oriented environment in which to study (not uninfluenced by the reduction in the age of majority and current economic pressures on the young), coupled with the more vocationally orientated attitudes of staff and other students, are seen to be attractive features of further education. However, the take-up of the available opportunities for part-time study has varied considerably in different parts of the country, and has ranged from good for boys employed in engineering, to poor for girls in the distributive trades. The Industrial Training Boards and various other government schemes have had varying success in encouraging continuing education. Further education has tended to be responsive to demands made upon it, as illustrated by the number of college-devised schemes which are available only on a local basis. The revision of technician courses under the Technician Education Council (TEC) and the Business Education Council (BEC) which supersede National Certificates and some City and Guilds schemes, has placed considerable responsibility on colleges to devise courses appropriate to local needs while still trying to preserve the merits of a nationally recognized qualification.

It is unfortunate, but nonetheless true, that the further education sector has in the past existed with scant attention to, or from, the schools. Successful colleges in the expansionist 1960s operated under an almost piratical ethos that students were prizes to be captured from rival institutions, whether schools or colleges. This was done by blandishment, although usually with the implicit and sincere belief that the student's interest was paramount. Large-scale planning was haphazard, and systematic study of the sector was notable by its absence.

Higher education
Higher education (usually defined as being concerned primarily with post-GCE A level) has been reorganized following the *Robbins Report* of 1965 [2]. A network of 30 polytechnics has been established as a public sector alternative to the universities. They offer many CNAA validated degrees (replacing earlier concentration on London external degrees) and HNDs, usually with

the option of full- or part-time study. The university sector of higher education has, to a large extent, been preoccupied with the problems of expansion in the 1960s and economic recession in the 1970s. It is, however, expected that by 1981 about 529 000 students will be involved in higher and advanced further education.

Nevertheless it would be wrong to assume that apart from increasing opportunities the universities and polytechnics have done nothing in the fields of teaching and learning and course design [12]. The potential higher education students of the eighties will not only have a wider choice of institutions and courses but may well, if current trends continue, represent a wider range of schools and social background than their counterparts in the fifties [13]. There are also clear indications that they may well be better taught during the degree course itself. Higher education, in common with primary and secondary education, is becoming more conscious of the educational needs of the individual and the cultural and economic needs of the community.

Teacher education

We would wish to draw particular attention to the reorganization of teacher education, which has had, and will continue to have, a marked effect on science education in the schools. The two traditions of science education discussed above were derived in part from the needs of the schools and in part from the sharp distinction between teacher training courses in the training colleges and those in university departments of education. The reconstitution of the training colleges in 1966 and the institution of the B.Ed. degree did much to force colleges of education towards an improvement in academic standards, sometimes at the expense of professional relevance. Economic and recruitment pressures in the late 1960s militated against science departments in the colleges of education and the recent rationalization of the colleges following the publication of the White Paper *Education: A Framework for Expansion* [14], which will reduce initial training places from 116 000 in 1971 to 36 000 in 1981/82, has led to the cessation of science teacher training in many institutions. By 1981/82 it is anticipated that only a relatively small percentage of the total planned output of science teachers will be trained outside the university sector, which still tends to provide a high level of training to graduates in the separate sciences, sometimes with a grammar school type of academic curriculum as the focus for professional studies. Reorganization has furthermore forced some colleges into difficult marriages with polytechnics lacking a traditional involvement in teacher training, and left others as diversified institutions lacking a traditional involvement in anything but teacher education.

In summarizing this review we would wish to note the wide variety of educational provision at primary and secondary levels—a range of provision that is often criticized as being uniform and conformist—and the greatly increased opportunities that now exist for further and higher education. In particular we note that while most of our analysis has been concerned with educational provision in England and Wales, science education in Scotland has during the same period

17

also undergone significant change. The Scottish Education Department, like the Schools Council, has undertaken major programmes of curricular reform in order to meet changing educational needs. The main difference between the two systems is in the degree of central control rather than in the content of the curriculum. We have, in this section, refrained from extensive comment on the factors that have led to this change or, apart from our strictures in the area of teacher training, the benefits or otherwise of change. Our concern here is to analyse rather than judge, in the sense that irrespective of one's views as to the merits or demerits of the tripartite system of the 1950s, it is patently not the system of the 1980s. We do however note that the movement towards comprehensive secondary education and increased opportunity for higher education was a clear political trend of the early sixties that the school curriculum movement did not reflect: as the map of school science was redrawn, a superstructure of science curriculum development was created in order to satisfy traditional aims and aspirations in an educational context that was accommodating comprehensive reorganization. Whether that same system has to any significant extent come to terms with the concomitant educational and ideological changes that are implicit in the comprehensive movement is the concern of the remaining sections of this part of the report.

SOCIAL CONTEXT

Introduction
Any discussion of educational issues and priorities has to be placed in its social context; although what follows is no more than an outline, it does indicate the social trends that will influence the decisions to be made about science education. We must recognize the existence of opposing forces—those concerned on the one hand with the maintenance of the *status quo,* and on the other with redefinition of the whole basis of our society. There are those who perceive, for instance, a society in which cherished values and standards are threatened, and others who, from the opposite point of view, believe that there is an establishment protecting its own interests and opposing change. It would be difficult to defend either of these extreme views. That there is a tension between the opposing influences which they may generate is scarcely a matter for argument. The variations in opinion, which this document can in no way resolve, concern the nature and degree of change that has already taken place, and the extent to which particular changes may be regarded as real, superficial or, even, illusory.

There are two points of view. Those who believe the changes to be real see a shift from a class-based society, with well-defined norms and values and little social mobility, towards a pluralist society in which individual and group values are negotiated rather than imposed. Traditional lines of authority, whether they be in the factory, the school or the family, have been eroded and are being replaced by complex systems of temporary relationships arrived at through expediency or from a redefinition of the role of the individual in society.

According to this belief the roles of teachers, parents and pupils are as open to redefinition, as are those of ethnic minorities, shop-floor workers or managers. The school is becoming an open campus that serves, and is part of, the community at large [15]. Teachers whose judgements were rarely questioned are becoming accountable to governors, parents and employers. Pupils are becoming increasing involved in decision-making. This situation demands a greater degree of open discussion. A consequence of this is that science would become an open activity, sensitive to the wide range of social needs.

Those who believe the changes to be illusory think that our society is not moving away from class divisions towards pluralism. In their view effective control rests with small, relatively static groups whose influence stems from the ownership of property and wealth, and from the tenure of positions of legislative and administrative power. They would concede that many more issues are apparently open to public debate, but would argue that the outcome is rarely influenced by those outside the "establishment".

According to this second belief the educational system is a tool of the class structure and functions to perpetuate the system through selection and grading. It provides the basis for an industry in which a relatively small number of highly educated scientists and technologists monopolize the skilled creative activities, a somewhat larger number of technicians carry out skilled work which requires some scientific/technical background, while the vast majority do jobs which require very limited skills, and little scientific or technological background [16a].

Schools are expected to grade people for different kinds of jobs and to educate them appropriately [16b]. This, combined with the effect of higher education, means that the main pressures on secondary education are for the scientific education of only a minority, contrary to the true interests of the majority.

To consider only two models from a wide continuum of views is to risk charges of over-simplification. However, they represent points of reference between which individuals may identify their own positions.

Implications
At the school level, perhaps the most important single factor that arises from the above analysis is the need for an open discussion of purposes. In the absence of open debate the social changes we have described result in a tension between the old and the new that can easily lead to open conflict as old values are imposed by one party and rejected by the other. Where discussion is built into decision-making processes the possibility exists, if it is not always grasped, for old and new values to be synthesized, a process that can lead to creative development. It could be argued that many of the behavioural and motivational problems encountered in our schools arise not so much from the traditional problems of adolescence as from the failure of teachers, parents and pupils to renegotiate value positions in a pluralistic and changing society.

One consequence of rapid social change that has strong implications for education is the radical shift that has taken place with respect to the "image" of education and its economic base. The curriculum development movement of the early 1960s was related to the following assumptions:

(a) that the period of formal schooling would be extended to 16+;
(b) that the number of pupils in our secondary schools would continue to rise;
(c) that the provision of places in further and higher education would continue to expand;
(d) that our economy would continue to be based on rapid industrial growth and an increasing Gross National Product; and
(e) that more, and better, educational provision would lead to a fairer society and more equal distribution of wealth.

Education, and its general expansion, was seen both as a social and economic "good" directed to both qualitative and quantitative ends. Since the mid-1960s these assumptions have been either questioned or proved false, such that it is now widely recognized that education does not have a direct influence on the economic prosperity of the country. Indeed, it may have a negative effect, in sheer economic terms, by channelling the highly educated (and therefore expensive) personnel into non-productive jobs. This means that the educational system must expect to give a clearer justification for its activities in the future if cuts in expenditure are to be avoided. The raising of the school leaving age in 1973 has now to be seen in the context of the sharp drop in the birth rate which is resulting in a longer school life for fewer pupils. For economic and political reasons the expansion of higher education has been curbed, although there is clear evidence that further education will expand to order to counterbalance youth unemployment. The energy and monetary crisis of the mid-1970s led to the consideration of zero or negative growth economic models and the general stagnation of productive industry. The increasing awareness of the finite nature of energy and material resources and the problem of the developing nations may, or may not, lead to the careful and highly controlled utilization of new resources.

Finally, society has been slowly, and perhaps painfully, coming to the conclusion that improved educational provision does not *in itself* lead to social change, and that science *by itself* cannot solve all the problems. This has resulted in a crisis of confidence in the effectiveness of our educational system and a strongly overt pressure for greater public accountability [17]. While it may be too early to accommodate all the above changes, it is clear that the educational system will, over the next few years, undergo a substantial review and reappraisal. The key issue is not that the review will take place, but who will undertake it and establish the appropriate evaluative criteria. This, more than any other single factor, represents the immediate challenge to the teaching profession and the rationale for our attempt to review the contribution science teachers can make to educational programmes in the future.

THEORETICAL CONTEXT

Introduction
We suggested earlier that a comparison between the grammar and modern school traditions of science teaching pointed up important issues regarding the

nature of scientific knowledge and the related problems of teaching and learning. In this section we wish to draw attention to a number of significant changes that have occurred since the publication of the last ASE Policy Statement.

Science teachers have not, until quite recently, subscribed to *explicit* psychological or sociological models when devising teaching and learning schemes. Nevertheless, science teaching has always been based on a range of implicit values and assumptions concerning the psychological development of young people, the nature of learning, the nature of knowledge in general and scientific knowledge in particular. This section of our report is largely concerned with exploring these assumptions in the context of recent theoretical studies.

Traditional school science teaching, with its emphasis on the textbook and chalk-and-talk methods, accepted as given a generalized stimulus–response psychological model; that is, *all* the pupils of a given age and ability range would respond in a more or less similar way to the intellectual stimulus of the teacher. These assumptions were revealed in the formal design and layout of teaching laboratories, carefully selected and graded textbooks, and an emphasis on the selection and rigid streaming of pupils. Equally important, it was assumed that pupil ability was largely an unvariable factor, that the total pool of ability was limited and fixed, and that the content of school science was unchanging and non-problematic. Scientific attainment depended entirely on matching groups of pupils to a course content appropriate to their level of intelligence and ensuring adequate motivation. Motivational issues, if discussed at all, relied heavily on a system of rewards and punishment and a combination of the Protestant ethic and values relating to deferred gratification, and pedagogy was dominated by behavioural psychology [18].

Science teachers who rejected these traditions relied heavily on a combination of lecture demonstration and class practical work as a means of transmitting the same scientific content. Two dominant theoretical assumptions underpinned this alternative strategy: that learning by doing was more effective than learning by listening; and that learning by doing was more inherently interesting for the pupils [19]. It is important to note that neither of these alternative rationales necessarily altered the content of teaching programmes, for too often the dictated note to 30 able youngsters was replaced by a common practical exercise or a common project.

Secondary modern science teaching in the main replicated the assumptions and procedures of the grammar schools but operated on a more general and reduced content. Some teachers had, however, developed a very different rationale influenced by the work of Dewey, Piaget and others [20]. These teachers placed a heavy emphasis on pupil-initiated project work, question-and-answer teaching techniques, and a study of the science of everyday things. Learning by doing was linked to a pedagogy that dealt with the concrete before moving to the abstract and relied on motivational factors such as interest, relevance, and a consideration of the applications of science.

21

The curriculum

The science curriculum development projects of the 1960s have, in their different ways, espoused aspects of these earlier traditions, although only Science 5/13 and SCISP have developed teaching and learning strategies based on explicit psychological models [21]. The major Nuffield projects have embraced pupil practical work and guided heurism linked to a partial acceptance of Piagetian stages of intellectual development. Nuffield Junior Science and Secondary Science lend themselves to project approaches which attempt to relate school science to everyday life. SCISP has extended science studies to, and across, the interface with the social sciences. A major weakness, however, of all these developments is the failure to move away from the same behaviourist assumptions that characterized science teaching in the previous decade. Content is still largely prescribed, and access to knowledge controlled by the teacher; group teaching methods predominate; and assumptions regarding the innate and fixed abilities of young people are written into the various teaching programmes. From a theoretical standpoint a major criticism of Nuffield and Schools Council developments is their failure to acknowledge the psychology of individual differences. Private enterprise, at the cottage industry level, has made some attempt to acknowledge this important theoretical shift through the development of science teaching programmes utilizing resource-based, independent or individualized-learning strategies. In general these approaches to the science curriculum have been empirical responses to changes in the educational context arising from comprehensive reorganization, and they have been motivated by organizational changes, such as de-streaming, mixed ability teaching groups and the rationalization of the timetable.

In other words, recent developments have been motivated by a need to respond to organizational changes and have not arisen through a rethinking of theoretical positions. Thus recent criticisms of Piagetian developmental psychology [22] have not been assimilated into curriculum thinking and there is still an almost blind allegiance to the dated concept of development stages. In many ways the old orthodoxy of fixed abilities has been replaced by fixed stages and the notion that some youngsters never reach the level of formal operational thinking. Such a rationale also provides an excuse for failing to evaluate the effectiveness of teaching and learning programmes, as did earlier arguments based on a genetically-determined intelligence factor. It is, moreover, highly significant that to date there is little evidence of the development of an alternative teaching strategy based on psychological models which place a high premium on individual interpretations of meaning and the role of language in mediating the individual's growing understanding of the world. It is probable, however, that these non-normative and personalized approaches to human behaviour will exert an effect on educational thinking in due course [23].

The nature of knowledge

School science has traditionally been regarded as a fixed body of knowledge, related to and derived from "real" science, which young people need to acquire in order to understand the world they live in and which they must master in order to become scientists. The curriculum reforms of the 1960s reiterated these assumptions when the map of school science was redrawn within similar boundary lines as before and one item of content gave way to another. While, with varying degrees of emphasis and conviction, the writers of the Nuffield courses stressed the illustrative nature of the course content and their hope that teachers would adapt programmes to meet their own needs, it should be noted that a combination of the examination system, parental pressure, professional inertia and commercial interests, has led to a second generation of teaching schemes which remain highly codified. Secondary school science is still firmly characterized as being fixed and non-negotiable in contrast to many other areas of the curriculum. The definition of science in terms of its method rather than by the nature of its, subject matter has to date had little effect on the school curriculum, especially in terms of any attempt to integrate science with other curriculum areas. Such changes in content that have occurred have, with notable exceptions [25], reduced the emphasis to be placed on applications and social relevance, and tightened the boundaries between the individual sciences, and between the sciences and other forms of knowledge. As a result, secondary school science has become steadily more isolated from the totality of the school curriculum.

These trends in defining the content of school science courses run directly counter to one set of social changes outlined in the previous section. They fail, furthermore, to take into account current debates concerning the nature of knowledge, the form and structure of school knowledge and the nature of the scientific process itself. The forms of knowledge enunciated by Hirst and Peters [26] allow science to adopt a highly élitist position, a position reinforced in the school context by the relatively high level of physical and manpower resources allocated to science. In Berstein's terminology [27] science is tightly framed and is easily able to resist attempts either to popularize it or to integrate it with other curricular activities. Hence applied science, craft science, technology and the various formulations of general or integrated science are readily demonstrated to be "inferior" in both intellectual and vocational senses. The problem is compounded by much modern writing on school science education which emphasizes that while content is, and always will be, important, the main thrust of modern programmes is in the direction of teaching scientific methodology and processes—a methodology distinctive to the science and powerful in its range and applicability. As has been stated elsewhere [28],

Methods in inquiry used in science, commonly called "scientific method", are intrinsically elusive and difficult for the layman to grasp. It is a method neither in the sense of a formal procedure nor an infallible prescription; rather, it is a set of attitudes, springing from the philosophy of the discipline, which provide a basis for action.

This contrasts sharply with the almost universally agreed and accepted formulation of scientific method as taught in schools; a method that is formal and consisting of set procedures that can be systematically applied to the resolution of

23

any problem arising in the course of a normal and well planned lesson. The Kuhn–Popper debate [29] does not appear to disturb the calm methodological seas of school science.

Most science teachers, who are themselves products of a science education that places a high premium on scientific knowledge and pays lip service to the history and philosophy of science, share with many practising scientists a scant understanding of the nature of scientific knowledge itself. They see science as a body of knowledge, a methodology and a set of values that exist in their own right, uninfluenced by the social world. They fail to see that ordinary human values are applied, and misapplied, in the selection of scientific problems, the resolution of methodological issues and the preparation and presentation of results. Science appears to exist outside any valid social context. It is objective, value-free, and totally aseptic. The popularizers and sceptics, such as Bronowski and Koestler, are dismissed almost out of hand, and the writings of Medawar and Jevons are regarded, if considered at all, as mild perturbations of the otherwise equable climate of scientific theory and practice. Any attempt to relate science to politics, or technology to social change, is resisted on the grounds that knowledge is knowledge, and application or utilization is somebody else's concern. Science, in spite of its humanistic traditions, has become a slave to the high technology society and is more concerned with system maintenance than social purpose. At the school level this bias is reflected through syllabuses which emphasize theory and reduce the opportunity for real personal involvement. As a result the pupil is presented with a set of findings established with the aid of a non-problematic methodology. At no point is the pupil required to negotiate his or her own meaning or explore the real problems of science as it *is*, rather than as it is written.

A lack of opportunity to explore the history and philosophy of science and to study science in its social, economic and political contexts is perhaps a major contributory factor to the popular image of science gained by young people. What they see is a subject dominated by facts and principles established outside their experience of everyday life. They study a subject isolated from its history or context and which is also very demanding, time-consuming and complex. This image results in part from the failure to locate science studies in the everyday context of life and work, and in part from the failure to integrate science into the common culture. The majority of young people fail to de-mystify the subject and do not see science for what it is—one of the most important *cultural* activities devised by man.

Part Two   The Current Curriculum

INTRODUCTION

The major policy documents published in 1957 [30] and 1961 [31] were produced within a curriculum context which was more clearly defined than that of the present day. For those pupils in grammar and technical schools preparation for GCE O- and A-level examinations dominated the approach to the provision for science education. Given that slight variations existed between the syllabuses of the different examination boards, and also that differing emphases were placed according to the subject, it can be said that scientific knowledge was prescribed in terms of both intent and outcome. Apart from implicit assumptions about the use of rigour and logic there was no overall view of the function of scientific method in the aims of teaching; some teachers adopted a historically sequential approach and others used a form of logical approach which often became evident to the pupils only with hindsight. Practical work was almost entirely confirmatory with little opportunity for open-ended experiments—a situation influenced by the demands of the practical requirements of the examination boards. The major curriculum emphasis was therefore on content which often represented the body of knowledge required for successful progression to higher levels, and which in turn demanded a disciplined, teacher-controlled approach. The courses were specifically and purposefully designed for the top of the ability range, with a major emphasis in design being the expectations and interests of higher education.

Although industry clearly accepted an O-level certification as proof of ability and knowledge acquired, little attempt was made in the design of syllabuses to meet the needs of industry even in the very few instances where these were articulated. The linguistic, conceptual and numeracy content of syllabuses and courses were all demanding, although the selective nature of the pupil population often resulted in such issues becoming hidden until a much later stage in the school, when many would have been turned away from science. However, within the context in which such courses were placed the objectives were achieved to a very high degree. The external examinations system for GCE O- and A-levels ensured a degree of "objectivity" which generated confidence within higher education and industry, and which also created a feeling of security for teachers.

EXAMINATIONS AND THE CURRICULUM

As has already been observed (in Part One of this document), this description of the teaching of science at the secondary school level was not representative of the courses which were devised for pupils in secondary modern schools. There, little emphasis was placed on content or process, but the chief concern lay with the pedagogical problems of generating some understanding of science on the part of

25

the pupils. It was as a result of the Beloe Report of 1960 that the Certificate of Secondary Education was introduced in 1965 to provide an examination for candidates who had completed five years of secondary education. The intention was to cater for those of around average ability, or above, for whom GCE O level was an inappropriate aim; reference was made to a group falling between the twentieth and sixtieth percentiles. A major feature of the CSE system was the variety of modes of examination which were offered, ranging from Mode I (an external examination based upon a syllabus prepared by one of the regional examining boards), through Mode II (an external examination based upon a syllabus submitted to the board by an individual school or group of schools), to Mode III (an internal examination, moderated and graded by the board, but set and marked within the school(s) responsible for devising the syllabus). The last of these (Mode III) was foreshadowed in the Beloe recommendation that "the teachers in their schools using the examinations must have a major role in operating them and shaping their policy [32]". Thus the CSE seemed to hold out the possibility, for secondary modern schools particularly, of maintaining a teacher-controlled curriculum with the advantages of externally-validated accreditation. In the event, Mode I accounts for more than 70 per cent of all subject entries and Mode III for 25 per cent (although the variation among boards is considerable; in 1974 the variation extended from 69 per cent Mode III in one board to 7 per cent in another). The pattern for science entries is not markedly different from the pattern for all subject entries, and it is interesting that Mode II is seldom used. Many of the Mode I science syllabuses resemble their GCE O-level counterparts, and examples of real originality are rare, for the prevailing emphasis in such syllabuses is more academic than vocational or technical. This arises because of the use, in some schools, of the CSE examination as a form of insurance for the weaker GCE candidate, or the need to create common sylla-buses in years 4 and 5 of the secondary school either to conserve resources or to delay the selection of final examination objectives. It may have been anticipated that, with the raising of the school-leaving age to 16, the CSE examination would have become more closely related to the needs of those going directly from school to work. In practice, the one major change consequent upon RoSLA was the extension of CSE examining to the lower ability end of the school population for which it was not originally intended. This was largely accomplished by the development of Mode III syllabuses, often with a limited grade range, and this has, to some extent, undermined confidence in standards. As a result examination boards have begun to press for the closure of limited range schemes, and the general regularization of Mode III. CSE is, however, a teacher-controlled exam-ination, and teachers themselves are responsible for setting limits on the flexibil-ity which they rate so highly. The curriculum developments of the sixties and seventies, linked to the possibilities of Mode III assessment, seemed to offer an unprecedented opportunity for teachers to influence much more directly the science curriculum of the schools in which they teach, an opportunity which at first sight appears not to have been taken to any significant extent. The reasons for this will be explored later.

THE SCIENCE CURRICULUM

The curriculum development work of the past 20 years has arisen from a variety of influences, among them being the policy documents of 1957 and 1961 of the Association for Science Education, the curriculum development movement in North America, and the establishment of two important funding agencies—the Nuffield Foundation and the Schools Council. The part played by the Association for Science Education in first of all formulating proposals for a renewal of the science courses in schools, and later (through its membership) in developing such proposals through trials schemes in the schools, has been repeatedly acknowledged. We would wish to endorse the valuable contribution made by the Association at that stage. However, what *began* as an attempt to provide continuous renewal and flexibility in the teaching of science has often emerged as a new orthodoxy. Thus "sample schemes" produced by many of the curriculum projects have been transformed into "established courses". Nevertheless, the movement has been important for the stimulus it has given to new thinking in the teaching profession, and for the provision of equipment and resources in the teaching of science. The new schemes have encouraged the exchange of information of ideas between colleagues; the training of the profession through both initial and in-service courses; and have attracted the attention of educationists overseas, particularly in Europe and the developing countries. It is the case that innovation in science teaching has been more readily achieved in Scotland than in England and Wales, partly due to the more centralized organization and assessment of courses. The general influence upon the nature and content of public examinations in England and Wales, and upon the institutions of higher education, has been more gradual. There are important respects in which the science syllabuses of the examination boards have changed little over the past two decades, but it could not be denied that the curriculum development movement has affected to a marked degree the external examination system and the manner in which science is taught in schools.

THE CURRICULUM IN DETAIL

No account of the present situation with respect to science teaching would be complete without an analysis of the curriculum developments which have taken place in the recent past. Such an analysis could take the form of separate reviews of each project, but we feel it preferable at this point to identify a few themes and approaches and to consider the more general assumptions which these have embraced. It is recognized that such an analysis is a highly reductionist process and, however helpful it may be for our present purposes, there remains a strong possibility that the major *interactions* between the various issues will not have been isolated.

In the analysis several issues arising out of Part One of this report have been identified and then used as criteria for the analysis of a variety of schemes. Neither the list of criteria nor the range of schemes chosen for inspection is exhaustive,

	SCHEMES			
CRITERIA	Nuffield Junior Science	Science 5–13	Nuffield Combined Science	Scottish Integrated Science 1st cycle
1 Age range	5–(12)	5–13	11–13	12–14 (S) 11–13 (E)
2 Ability range designed for	*All*	*All*	*All*	*All*
T: top/*M*: middle/*L*: low	*All*	*All*	*All*	*All*
3 Scientific knowledge	*N*	*N*	*P*	*P*
P: prescribed/*N*: negotiable	*N*	*N*	*P*	*P*
4 Scientific method	*Experi-*	*HD*	*HD*	*Neither*
I: inductive/*HD*: hypothetico-deductive	*menting*	*HD*	*HD*	
5 Practical	*D*	*D*	*D*	*C*
D: discover/*C*: confirm	*D*	*D*	*C(D)*	*C*
6 Emphasis	*M*	*M*	*M(C)*	*C*
C: content/*M*: method	*M*	*M*	*C(M)*	*C*
7 Teaching methods	*D*	*D*	*D(T)*	*T(D)*
D: discipline/*T*: teacher/*P*: pupil	*P*	*P*	*D(T)*	*T(D)*
8 (a) Choice	*P+T*	*P+T*	*T(P)*	*None*
P: pupil/*T*: teacher controlled	*CMR*	*CMR*	*CR*	
in (b) *C*: content/*M*: method/*R*: route	*P+T*	*P+T*	*T*	*None*
	CMR	*CMR*	*CR*	
9 Connections:	*SETDO*	*SETDO*	*E*	*Few*
*S.E.M.T.D.Ec.O.**	*SETDO*	*SETDO*	*E*	*Few*
10 Linguistic demands	*NO PUPIL MATERIAL*		*High*	*High*
11 Numeracy demands	*NO PUPIL MATERIAL*		*High where such exists*	*Low*
12 Conceptual demands *C*: concrete/*A*: abstract/*HA*: highly abstract	*C*	*C(+A)*	*(C+)HA*	*C+A*
13 Resource implications:				
Teacher in-service training	*ESSENTIAL*		*Yes*	*Yes*
Technician preparation	*No*	*No*	*Yes*	*Yes*
Costly equipment	*No*	*No*	*Yes*	*Yes*
Library resources	*Yes*	*Yes*	*?*	*No*
14 Prior knowledge	*None*	*None*	*None*	*None*
15 Terminal – *Yes/No*	*No*	*No*	*No — leads to 'O' level*	*No*
16 Role of external examination	*N/App*	*N/App*	*N/App*	*N/App*

Table 2

* *Key – S* =Social; *E* = Environment; *M* = Employment; *T* = Technology; *D* = Decision making; *Ec* = Economic development; *O* = Other school subjects (not English and Maths.)
† Dependent on topics and routes.

Nuffield 'O' Level	S.C.I.S.P.	C.E.S.I.S.	Nuffield Secondary Science	L.A.M.P. 1st cycle	Working with Science	Nuffield 'A' Levels		
11–16	13–16	13–16	13–16	13–16	16+	16+		1
Top 20–25%	Top 20–25%	$M \rightarrow L$	M	All	M	T	Intention	2
$T \rightarrow L$	$T \rightarrow L$	$M \rightarrow L$	$M \rightarrow L$	$M \rightarrow L$	$M \rightarrow T$	T	Outcome	
P	P	P	P	P	P	P	Intention	3
P	P	P	P	P	P	P	Outcome	
HD	I	I	I	Neither	Neither	HD	Intention	4
HD(I)	I	I	I			HD	Outcome	
D	D	C	D(C)	C	D(C)	D(C)	Intention	5
C	C(H)	C	C	C	D(C)	C(D)	Outcome	
M(C)	M(C)	C	C	C(M)	C(M)	C(M)	Intention	6
C	C(M)	C	C	C(M)	C(M)	C(M)	Outcome	
D(P)	D(P)	D(P)	D(T)	D(T)	P(D)	T(D)	Intention	7
D(T)	D(T)	D	D(T)	D(T)	P+D+T	D	Outcome	
P(T)	T	T	T	T(P)	P	V. little	Intention	8
	CR	R	R	C	C			
T	T	T	T	T(P)	T(P)	V. little	Outcome	
R	R	R	R	C	C			
SET	SETDEc	SETDEc	ET	†	All	SETEc	Intention	9
Few	SETD	SET	ET		Varied	Few	Outcome	
V. high	V. high	Mod.	Mod.–High	Mod.	High	High		10
V. high	V. high	Mod.	Low	Low	Mod.	Varied		11
(C+)HA	(C+)HA	C+A	C+A	C+A	C+A	(C+)HA		12
Yes	Essential	Yes	Yes	No	Yes	Yes		13
Yes	Yes	Yes	Yes	Yes	Yes	Yes		
Yes	Yes	Yes	Some	No	No	Yes		
Yes	Yes	Yes	No	No	Yes	Yes		
None unless following from Combined Science			None	None	Some	'O' level		14
Yes	Yes	Yes	Yes	Yes	Yes	Yes		15
Emphasizes content	Emphasizes content devalues wider aims	Varies with CSE Mode III			N/App	Emphasizes content		16

although we do believe that both are sufficiently representative for general conclusions to be made. The results of applying the criteria to the schemes as published and commonly taught are summarized in Table 2.

Reference to Table 2 leads to several points concerning the criteria considered.

(a) *What view of scientific knowledge has been assumed?*
It is probably a fair generalization to suggest that the lower the age group and the less academic the pupil, the less closely prescribed is the content of scientific knowledge for most schemes. The Schools Council Science 5/13 and Nuffield Junior Science schemes, for instance, offer considerable flexibility in both content and approach. The Nuffield A-level Chemistry scheme, on the other hand, leads to an examination which is based very firmly on the published materials, so that although the teaching order and the teaching methods may vary a great deal from school to school, there is little scope for deviation from the specified content. Although many of the projects draw attention to the steady evolution of scientific knowledge and understanding, most of them prescribe fairly closely both the body of information to be learned, and the nature of the "thought models" to be adopted. The absence of any dynamic can easily convey the impression of science as a mass of enshrined data and hypotheses. Many projects carry a high level of prescription and because of this the overall image is that, in this respect at least, there as been little change in direction as a result of the curriculum movement. This prescription of content may be the result of the demands of external assessment, although it is interesting to note the extent of prescribed knowledge even for those projects which do not lead directly to a terminal examination.

(b) *What view of scientific method is assumed?*
This analysis distinguishes between an *inductive* approach (which is essentially the process of reasoning from particular cases to general conclusions) and a *hypothetico-deductive* approach (in which specific conclusions are drawn from previously generated hypotheses). There are projects in which the pupils are specifically taught to gather data and generate hypotheses (Nuffield O-level Biology is one example) because "this is what scientists do". Others, by historical illustration, recognize that developments are frequently haphazard. Often the scientific method is ill-defined and most projects seem to avoid committing the pupil to any particular view of methodology with the consequent danger that there is created a false impression that the orderly framework in which the subject is presented reflects the manner in which the knowledge was first obtained.

(c) *What is the function of practical and experimental activities?*
Practical work in traditional courses was frequently illustrative or confirmatory. In many respects little has changed. True experimentation and discovery are encouraged by the pre-secondary schemes, but in those primary schools where science has long been taught, this may have always been the case. There is a genuinely experimental investigation in the Nuffield A-level scheme, and courses have project work of a truly investigatory type (Nuffield A-level Biology and Nuffield A-level Physical Science, for example). Other examples are not easy to

find however, and many of the Nuffield courses have spawned contrivances designed to produce the "right" answer nearly every time. The pupil is therefore lulled into the belief that he has "discovered" something as a result of turning a switch to start a device which cannot fail. This is not to deny the value of illustrative practical work, nor to suggest that true "discovery" is the only effective learning method, but simply a plea that we should make explicit the function of the experiments which the pupils are required to conduct. There remains a place for the so-called "guided discovery" (sometimes referred to as "stage-managed heurism") but it should be acknowledged for what it is.

(d) *What is the relative importance of content and method?*
We have already noted that most examination courses remain content-orientated, and even those courses which do not lead directly to a terminal examination are heavily influenced by the content. With the exception of the Schools Council *Science 5/13* and *Working with Science* schemes, any choice of content is restricted to teacher choice of route. However it must be noted that genuine advances have been made in assessing skills other than factual recall, and it is now fairly common practice for examination boards to state the weighting of test marks for a range of cognitive and practical skills, and even for some affective skills.

(e) *What assumptions are made about teaching method?*
The shift from the idea of the teacher as the *sole* arbiter of the choice of material to be taught is one of the most significant movements to be observed. There is a tendency for increased pupil participation in many of the more recent projects, such as the *Computer Assisted Learning Project* and the *Independent Learning in Science Project,* and examples of this change are to be found in almost all the projects of the last 20 years.

(f) *What developments in language, concept and number are assumed?*
Conscious attempts to develop basic language and number skills are almost entirely absent from the science teaching projects of the period under consideration. It is true that the encouragement of pupil/pupil and pupil/teacher dialogue will have an effect on oral skills, and the reduced "formality" in the recording of practical work has refined pupils' understanding by calling upon them to articulate their own impressions. Many teachers are not equipped to exploit these potential advantages, however, and much remains to be done. In reducing the emphasis on numeracy, as many schemes appear to have done, an important opportunity has been lost. It was wise to remove numerical barriers to understanding, but are there not situations in which science can provide both the example and the motivation for the improvement of "numeracy"? Linguistic, conceptual and numeracy demands are consistently high for O-level courses, and are often quite high for courses designed for pupils of average and below average ability.

One development which has departed radically from many of the assumptions underlying the projects reviewed above is the *Independent Learning in Science*

31

Project (ILIS). Although the interpretation of what constitutes an independent learning programme varies widely among the teachers who subscribe to the project, it would be true to say that in every case the pupil is placed in a situation where he is required to make independent decisions as a result of working through a planned series of tasks on his own. In some cases this may involve the use of a mechanical aid (for example, a tape recorder) but it may also involve contributing a piece of individual work in a largely teacher-led course. Such courses can be highly prescriptive, but attempts are always made to encourage work initiated by pupils. No assumptions are made about content, previous knowledge or ability of individual pupils, and the principal aim is to promote efficient learning through the optimum use of all available resources.

In the further education sector, the development of curricula has not been a coordinated activity. With the establishment of the Technician Education Council (TEC) an administrative structure has been imposed across the various institutions, but curricular decisions are firmly placed with individual colleges. The administration requires that syllabuses should be written in learning objective form, and that they should be designed to meet the needs of local industry (to the extent that evidence of consultation with industry is required). Bloom's taxonomy of educational objectives is required as the basis of assessment schemes, but the relative weightings given are largely at the discretion of the colleges. The content of the TEC courses is entirely negotiable with no philosophical view of scientific knowledge prescribed. The teaching method is determined solely by the teacher, and in its practical aspects emphasizes a reinforcement of theoretical principles together with the development of experimental skills with instruments and in other specific techniques. The courses are modular in nature and are devised in consultation with industry, and there exists a large degree of freedom in choice of teaching sequence. The resource implications for TEC are very serious. As each college must devise its own programmes and syllabuses, the teacher time involved is substantial. In-course assessment places an extra burden in time and expertise, and the range of skills now demanded of the further education teacher is very wide. All of these factors call for an extensive programme of in-service training in the immediate future.

SCHOOL-BASED DEVELOPMENTS

The uptake of the published curriculum projects has been modest [33], but the influence which the movement has produced on the science syllabuses of the examination boards has been impressive. Part of the problem of the acceptability of the published material for school use has been the changing context of the schools themselves. Projects which were designed unambiguously for a selective school population, for example, do not transplant effectively into a comprehensive school. The reaction of science teachers has been to attempt to devise their own schemes better suited to the needs of their pupils, either from published sources or by developing material of their own. This has been, for so many teachers, a huge task and one which initial training and in-service courses

32

of even the recent past have hardly fitted them to tackle. Nevertheless, much has been accomplished and the ASE has also contributed to teacher-based developments [34].

In addition to their lack of expertise in curriculum development, all teachers have experienced widening expectations by school and society concerning their role as teachers. Thus science teachers have become year tutors, careers advisers, pastoral care supervisors, counsellors, and administrators, with a consequent increase in pressure due to the restriction of time for developing their own subject. There has also been a significant shift in the task orientations of the heads of science departments, which may be broadly identified as a move from the "instrumental function" (concerning the organization of material resources) to the "expressive function" (involving the organization of human resources, through the coordination of science staff into a team, for example) [35]. In view of these *extra-science* pressures it is surprising that science teachers have achieved so much in the generation of school-developed courses, and it has been possible only because of the support science teachers have received through the employment of technicians, media resource officers and the availability of audio-visual and reprographic apparatus. If the economic climate worsens, it will be more difficult for science teachers to maintain the present level of curriculum development; continued financial support is an urgent requirement in this respect.

The development of school-based innovation has also promoted the introduction of subjects into the science curriculum hitherto regarded as having only a tenuous connection with "pure" science. Thus the development of schemes for the introduction of technology, applied science and engineering has been supported strongly by industry and other agencies, although the impact of such schemes in the schools has been spasmodic. The validity of such subjects on the secondary school curriculum is questioned, and many teachers would maintain that they are more effectively introduced through the more "traditional" study of science. Sub jects such as geology and astronomy are also competing for places in the science curriculum, and controversy exists among science teachers concerning such issues.

IMPLICATIONS AND RESOURCES FOR THE CURRENT SCIENCE CURRICULUM

The above analysis has shown that a considerable range of both curricular material and approaches currently exists to support science teachers in the development of their teaching programmes. It is considered necessary at this point to present in some detail the nature of the resources teachers require in order to maintain current levels of provision and the achievement of the limited objective described.

Staffing
Following Houghton, schools have developed an organizational structure that restricts career opportunity. The primary school with its relatively low unit total

cannot designate particular subject responsibilities in a way possible in the secondary area. Consequently there is a vital part to be played by the head teacher in interesting and motivating non-specialist staff in the use of science as an integral part of the primary education [36].

The impressive structures built into the secondary school to support pastoral, academic and guidance faculties have led to specialist teachers finding themselves involved in new areas of pupil care. The resultant clash of job priorities has made the task of the teacher more onerous; in a climate of extensive curriculum development, the time available for teachers to develop their own curriculum materials has been considerably restricted. The increased range of teacher responsibility has made the job of head of department more managerial [37]. It is necessary to examine the possible changes to staffing strategies with the advent of falling rolls. We need to consider how the structure in science departments should adjust to an inevitable loss of manpower as the school contracts. The teacher–pupil ratio is a critical area that the Association should consider; strategies include smaller classes, "double manning" in a team-teaching structure, and a reduced teaching load. It should be borne in mind that the falling roll will affect all departments and that science will not be excluded.

The pupils

Class size has for many years been a talking point for teachers across the curriculum, and in England envious eyes are cast towards Scotland, especially in the science field. To state a particular number of pupils as a maximum in science classes is dangerous because maximum numbers are readily interpreted as minima, and it may be more realistic to state that the Scottish model has its advantages which England ought to move towards. There is scope for research into the various aspects of reduction in class size, including such questions as the following.

1. Is the sort of science taught in smaller classes different from that taught in larger classes?
2. Is the learning situation more effective in smaller classes?
3. Can smaller classes be shown to be more effective users of the present capital resources of buildings?

The Green Paper *Education in Schools, a consultative document* [38] and the HMI Working Paper *Curriculum 11—16* [39] both highlight the issue of girls in science and engineering. The Association needs to consider carefully the real lack of science opportunities for girls in many secondary schools. There are many factors which may have caused the very clear lack of girls studying science at all levels, and in formulating future curriculum changes particular attention needs to be paid to how the subject can be made more attractive to girls in an option area as well as in a core.

Teaching methods developed in the seventies have in part produced much controversy: mixed ability teaching, resource-based work and independent learning methods have all received some acclaim and some criticism. The

34

Association has in the past taken the view that it should acquaint science teachers with a variety of teaching methods [40, 41] without passing judgement on the effectiveness or efficiency of any particular method.

Subject content

The *Bullock Report* [42] encouraged teachers to consider language across the curriculum, and the Association has followed this by establishing a working party to examine language in science [43]. Similar moves are likely following the concern regarding mathematics in schools.

The notion of science across the curriculum has also given rise to debate within the Association. Further work would involve the Association in considerable consultation with other subject bodies, and although it does present problems the opportunity should not be allowed to pass.

The bringing together of science and technology has progressed mainly through the agency of the Standing Conference on School Science and Technology and, latterly, through the Science and Technology Regional Organizations. It must be said that if the Standing Conference had taken a more positive line of action and used funds in a more constructive way then science and technology and craft subjects would be much closer within the curriculum than they are at present.

Material provision

Resources made available to finance the curriculum developments of the 1970s and the raising of the school leaving age, allowed for expansion of school laboratory provision and equipment. This enabled a range of projects to be adopted in secondary schools on an unprecedented scale. In the future any new curriculum schemes in science will have to compete against other subject areas for restricted financial resources.

Inflation has had a dampening effect on science provision as the costs of apparatus, chemicals and materials have soared. Future science curricula need to have regard to the reality of inflation as well as to reduced provision. External agencies such as Shell, BP, Esso, ICI and Unilever have played their part in supplying teacher and pupil resource materials at what must be a fraction of cost. They need to be looked on as current resources which may be reduced in future as companies consider the value of their educational products in relation to their total commercial success.

The future of material provision in schools is open to considerable uncertainty and it is important that the Association makes explicit the criteria on which educational priorities should be based on the allocation of science resources. In setting the scene for future strategies the Association needs to state clearly that its policy has taken public accountability into consideration.

Schools and industry

For some 20 years there have been moves to bring the world of work closer to the world of education, with each new effort being greeted as a move that would

make work in school appear more relevant to the pupils. In 1952 the chairman of Ratal Ltd and British Plessier Ltd said "the aeronautical industry wants all-rounders; we do not want people specially trained . . . but with a good deal of sound knowledge and a healthy attitude to life [44]". Over the years the demand for a general education has become more vocal [45].
These demands fall into four categories.

1. To introduce some aspects of the world of work into the curriculum; this, it was said, would be achieved by using relevant amd more overtly industrial or commercial applications in the lesson material [46].
2. To improve children's appreciation of the world of industry, especially in the field of technology [47].
3. To improve workers' knowledge and their awareness of the world of work, especially in manufacturing industry [48].
4. To encourage teachers and industrialists to work together to produce teaching material that is based on present practice in schools and industry [49].

The Schools Council Working Paper No. 7 (1966), although aimed at careers education, indicated clearly that closer cooperation between teachers and industry was desirable in the development of a technological society [50]. In 1977 the Schools Council again emphasized the continuing need to bring schools and industry closer together by formally setting up the *Schools Industry Project*.

Although the moves towards relevance are now directed across the curriculum, science educators are being urged more than ever to include in their schemes of work and syllabuses a greater awareness of industry.

Looking to the future, there is an urgent need to scrutinize science curricula in order to consider their relevance in the changing industrial and social contexts. This exercise is necessarily a massive one and will require in-service training of present teachers and a change in the initial training programmes. It will require further efforts to be made to bring practising science teachers closer to industrialists.

In-service implications

Given the present resources, the provision of in-service training forms a major support for curriculum development. Present financial strictures have led to a reduction of in-service provision not only at teacher centres and science centres, but also in the level of secondment. The move towards school-based in-service work could be viewed either as an attempt to consolidate the present curriculum or simply as an economic expedient. The future shows that whatever changes occur in the structure of science curriculum, there will be a demand for more in-service work in a variety of modes.

Whatever conclusion may be drawn from the present context, or whether one anticipates a common core, a common system of examining at 16+ and a broader sixth-form curriculum with N and F subjects, there is a need to review the direction that science education is taking. If it is felt that more than a slight adjustment to the present curriculum is needed then we have to consider the future, and this leads us on to Part Three of this discussion document.

Part Three   Curriculum Proposals

Models are undeniably beautiful, and a man may justly be proud to be seen in their company. But they may have their hidden vices. The question is, after all, not only whether they are good to look at, but whether we can live happily with them [51].

The purpose of this section of the report is to present a range of possible curriculum alternatives for the consideration of science teachers, advisers and curriculum planners. Before proceeding to a detailed exposition of the advantages and disadvantages of each curriculum "model", we need to consider more formally the aims of science education and the part education through science can play in general education.

Science, and science-based or -related activities, clearly have a contribution to make to the curriculum in general, but the nature of the contribution science is able to make depends on the way scientific knowledge, skills and attitudes are defined, developed and deployed. A fundamental issue is the extent to which science studies are seen to serve subject-centred as against more generalized ends. Thus science studies can be defined in terms of the development of the subject—the continuation and extension of scientific ideas—with the following broad **aims** being the most relevant in terms of school science education:

(a) the acquisition and understanding of scientific knowledge, generalizations, principles and laws, gained through a systematic study and experience of aspects of the body of knowledge called science;
(b) the acquisition of a range of cognitive and psycho-motor skills and processes gained through the repeated involvement in scientific activities and procedures in the laboratory and the field;
(c) the utilization of scientific knowledge and processes in the pursuit of further knowledge and deeper understanding, leading to the ability to function autonomously in an area of science studies. This also involves the ability to communicate with others.

In broader curricular terms we can say that the achievement of the above aims enables the individual to

(d) gain a perspective, or "way of looking at the world" that complements and contrasts with other perspectives or methods of organizing knowledge and inquiry, *and without which the individual cannot achieve a balanced general education.*

In short, the primary justification, and therefore purpose, of science education is to foster and develop, as part of the general education of the individual, a scientific way of thinking, a basic knowledge of scientific ideas and an ability to communicate with others. The precise realization of this purpose depends essentially on the way science is defined in conceptual and methodological terms.

The above aims can also be regarded as being person centred, in the sense that through the development of an awareness and understanding of scientific ideas the individual contributes to his or her own intellectual development [52]. But science studies, especially when organized around group activities such as practical work and fieldwork that involve interpersonal communication, provide opportunities

(e) whereby youngsters can gain a sense of social meaning and identity as well as personal autonomy.

Therefore a good science education should seek to develop a range of intellectual skills and cognitive patterns which would help youngsters to handle the problems of growing up in, and integrating with, a society that is heavily dependent on scientific and technological knowledge and its utilization.

Finally it can be argued that science studies that include the history, philosophy and social studies of science provide opportunities for

(f) explaining, and therefore understanding, the nature of advanced technological societies, the complex interaction between science and society, and the contribution science makes to our cultural heritage.

Clearly aims of this last sort are more socially oriented and again their realization would represent a significant contribution to general education. They would, furthermore, enable science teachers to redress the balance of curricular activities in favour of the aspects of general scientific literacy and understanding of science as a cultural as well as instrumental activity that have been highlighted in our earlier analysis of the current approach to syllabus definition and implementation. This broader definition of the purposes of science education would also meet Lord Bullock's assertion [53] that

All I am sure of is that the more it is possible, legitimately, to move away from a monolithic, mechanistic, dehumanized image of science; to establish a view of it as a humane study, deeply concerned both with man and society; providing scope for imagination and compassion as well as observation and analysis; and calling, in those who succeed in it, for outstanding personal qualities, the easier it will be to overcome the sense of alienation which turns many young people away from it.

It is central to the subsequent arguments presented in this section to regard the above six aims, or purposes, of science education as being common for all pupils irrespective of social background and ability. We present them as the essential aims of science education for all.

The personal and social aims of science education can be further explored by considering the variety of **contexts** within which scientific knowledge can be deployed. These may be conveniently classified as follows:

(a) *Science as science:* The pursuit of scientific knowledge as an end in itself, as an intellectual activity leading to the creation of further scientific knowledge. In this context a science curriculum would seek to establish the essential founda-

38

tions upon which higher education would build in order to equip the individual to undertake scientific research and development.

(b) *Science as a cultural activity*: The more generalized pursuit of scientific knowledge including aspects of its history, philosophy, literature and social context to effect a greater understanding of the contribution science makes to society and the world of ideas.

(c) *Science and citizenship*: The development of an understanding and appreciation of scientific and technological knowledge to enable active participation in the processes of democratic decision-making, especially in areas relating to the utilization of scientific developments and their technological applications.

(d) *Science in the world of work*: The development of an understanding of the way in which scientific and technological ideas are used to maintain an economic surplus and, in particular, the use to which science and technology are put in specific industrial, commercial and social situations.

(e) *Science and leisure*: The appreciation that science and technology provide a basis for a wide range of leisure activities and pursuits, and the development of a creative knowledge and understanding in this area.

(f) *Science and survival*: The development of an understanding of the role of science and technology in human survival interpreted in the broadest sense, but including aspects of self-sufficiency, the careful use of resources and the implications of alternative technologies.

Each of the above can be regarded as appropriate and legitimate contexts within which scientific ideas can be explored and developed; these aims should also be regarded as components of a sound science education programme. The problem for the curriculum planner is that of ascribing values and priorities across them whilst seeking to ensure that each is properly represented in a coordinated and balanced manner.

To summarize: A science education programme should seek to reflect general and specific aims which are on the one hand related to the contribution science makes to personal intellectual growth and development, and on the other to the areas, or contexts, within which scientific knowledge is used and deployed. To achieve a balanced science education within the context of general education requires the consideration of all the above factors as being of comparable status and importance. Given the educational and social analysis presented in Part One of this report, we would argue that parity of esteem with respect to the above factors is an essential goal for future curriculum planning.

THE STRUCTURE OF THE SCIENCE CURRICULUM

In seeking to define a number of curriculum models we have become conscious of the variety of educational institutions and curricular assumptions that exist in the British system. We are also highly aware of the constraints that exist as to the legitimacy of varying educational goals; educational resources—time, money, equipment and facilities; and the expertise and motivation of teachers and pupils. Finally we acknowledge that it is necessary to move away from the overtly philosophical and conceptual analysis of aims previously presented and we attempt to reinterpret those ideas in terms of content, method and resources at the practical level of curriculum planning.

In presenting a range of alternative proposals we have decided to consider the curriculum in three phases related to approximate pupil *ages* rather than types of educational institutions (first, middle, secondary, schools/colleges). By adopting an age/phase structure we do not necessarily wish to accept any particular psychological model of human development or suggest that the phases are anything other than a useful way of organizing curriculum content within the total continuum of educational growth and development. Our proposals are therefore organized, but loosely framed, within three distinct phases; Phase 1 (5–11 years), Phase 2 (11–16 years) and Phase 3 (16–18 years).

Set alongside these three chronological phases we present three possible curriculum models which seek to achieve the aims of science education presented above within a number of the contexts described earlier.

It is important not to regard these models as being simple alternatives, any one of which could serve as the sole basis for curriculum planning, but to regard them as attempts to highlight potential alternatives to current practice. Nevertheless the models do represent attempts to move away from current definitions of the science curriculum in order to achieve a broader coverage of aims and contexts. We are, after all, concerned at this stage with guidelines for future curriculum development as distinct from curriculum development itself [54].

Finally, by way of introduction to the curriculum models, we wish to comment on, and perhaps resolve, the content-versus-process debate that hinders clarity of thinking on the content and nature of the science curriculum. In our models we have accepted that few teachers would wish to attempt to construct a science syllabus in purely content terms, i.e. work on the assumption that science can be process free. Similarly we assume that a content-free process-based syllabus is equally undesirable, if such a syllabus were possible. Although varying weightings and emphasis can be given to content and process, we suggest the following parameters in curriculum planning.

(a) The *range of content* to be included in the syllabus; covering both issues of depth of treatment and breadth of study and the general parameters set for the inclusion and exclusion of areas of scientific knowledge (for example, would the syllabus include geology? astronomy? psychology? ESP?).

(b) The extent to which scientific knowledge, however defined, is treated in isolation from other areas of knowledge; the problem of subject integration both within "the sciences" and across the curriculum.

(c) The extent to which scientific knowledge is perceived to be value free and separate from any consideration of social context.

40

(d) The extent to which scientific knowledge is, or is not, related to technological, practical and everyday phenomena and situations; the "pure", versus "applied" versus "practical" debate.

(e) The extent to which scientific knowledge is presented according to an implicit, or explicit, internal logic or epistemology, or as an area of useful knowledge to be used to solve problems.

(f) Finally, the extent to which scientific knowledge is presented as objective knowledge [55].

It is the varying balances and emphases that are possible within and between these parameters that will give different approaches their characteristic flavour and direct an approach towards an emphasis on some aims, or contexts, rather than others. We are basically arguing, therefore, that while all syllabuses will have content, and all will reflect process, they may present very different "images of science" depending on how the above parameters are related to the contextual factors discussed earlier. Thus, while a chemistry syllabus defined, and taught, in an abstract way with no reference to practical application, social relevance or other areas of curricular activity may achieve certain of the aims stated earlier in this section, it will not achieve others.

SCIENCE STUDIES IN PHASE 1

In considering science in the first phase we recommend the effective implementation in all schools of the approach to science studies detailed in the ASE Policy Statement *Science for the Under-Thirteens* [56]. We propose that all primary and, where appropriate, middle schools, establish as a matter of priority a science education policy across the curriculum based on the following essential characteristics:

(a) That what is undertaken under the heading of science "should arise out of the spontaneous interests of the children and should not be imposed upon them with the aim of laying foundations", in a formal sense, for future science studies.

(b) That elementary scientific ideas should be derived from the exploration of the immediate environment and should involve "the application of an attitude of enquiry and the establishment of *personal* patterns of understanding from first-hand experience".

(c) That pupils should be encouraged through *the careful management of their learning environment* to make emergent generalizations of a temporary nature, and be given the confidence to accept that these will have to be modified in the light of further experience.

(d) That pupils, through individual and small group work, should be encouraged to speculate freely and creatively on the nature of objects and phenomena. In particular, a strong emphasis should be placed on the pupils' talking about, and discussing, science and on encouraging the creative expression of personal meaning in their own language and through modes other than formal writing.

(e) That all scientific work should arise within the context of an integrated curriculum. Science should not be separately timetabled, or be taught in a specialist room. Existing course materials should be available in all classrooms as part of the general stock of accessible resources [57].

(f) Finally, scientific phenomena should be freely used as the starting point for a wide range of creative work, e.g. poetry, story writing, drama, painting, model making, as part of the process of establishing a confident and open approach to science studies.

The above ideas require teachers to react quickly and enthusiastically to opportunities to explore science with their pupils. It is also essential that schools concerned with the education of pupils in this first phase have adequate physical resources to undertake the type of work we recommend. In our view one of the main reasons for the low level of work in many schools is poor resource provision. A further major prerequisite is the raising of the level of awareness and confidence of teachers to handle scientific ideas effectively and we will consider the implications of this in the final section of this report.

It is important not to express a directive regarding what is, or is not, appropriate, in terms of content and processes, for inclusion in the science studies of Phase 1. We feel that the list of activities presented in the ASE report are useful as a guide to teachers but we would suggest that activities such as "fieldwork in school grounds and open country; the examination and sorting of 'finds'; finding out more about specimens and situations by magnifying, weighing, measuring, etc.; the maintenance and observation of living material; and the improvisation and construction of equipment [58]" are more important than an over-concern with formal experiments. In short, we would place the main emphasis on exploration and observation rather than empirical methodology. We would also strongly support the wider use of radio, television and audio-visual aids as a means of expanding the learning environment of the classroom.

The end result of an effective programme of science activities as outlined above should be the integration of science with the other disciplines that are explored in this first phase. The emphasis will be on developing the right attitudes to scientific work rather than acquiring a formal knowledge and understanding of any prescribed areas of subject content. This is not to say that by the end of this phase a pupil will not know any science, but merely that this individual store of knowledge and experience will be idiosyncratic and highly personal. We believe this is as it should be and would provide the basic interest and motivation for the more formal study of science in the later stages of general education.

SCIENCE STUDIES IN PHASES 2 AND 3

Given the above framework for the first phase of science studies, pupils entering Phases 2 and 3 will differ in innate ability, acquired knowledge, personal experience and levels of motivation, although we would hope that in terms of interest and motivation, general levels would be even higher than those currently encountered among pupils making the transition into secondary schooling. The basic problem in the secondary school is that of moving towards a common programme of scientific studies while at the same time allowing full attention to be paid to individual differences of ability and aspiration.

In formulating the following guidelines we have noted Bruner's proposition [59] that

Any idea or problem or body of knowledge can be presented in a form simple enough so that any particular learner can understand it in a recognizable form.

which provides, if nothing else, a *philosophical* basis for the argument that a programme of science education for all in the 11–16 age range should not be predicated on different courses designed for pupils of different abilities, although it should allow young people to reflect varying interests and longer term aspirations [60]. In attempting to move away from the weaknesses of current provision described in the earlier sections of this report, and which normally involve the creation of different courses for different ability groups, we present three models, each representing a distinctive evolutionary stage. Thus Model 1 represents a fairly marginal reorganization of current resources in an attempt to avoid certain of the main problems of premature specialization in the science curriculum, while at the same time introducing an element of broadening. Model 2 develops these ideas more fully by introducing a greater concern for the application of scientific knowledge, skills and processes to the contexts discussed in the introduction to this section. Model 3 completes the process by adding activities related to developing an appreciation of the personal and social dimensions of science studies. In terms of curriculum innovation and development these models represent short-, middle- and long-term goals respectively.

MODEL 1

Phase 1
Introductory science as detailed on pp. 41–42.

Phase 2
Years 1 and 2   *Basic Science Core*. This would consist of a two-year foundation course designed to act as a bridge between the experimential elements of Phase 1 and the more formal studies of Years 3–5 of Phase 2.

The course would be *either* based on the more general content of existing science courses [61] with an explicit concern for the practical applications of that content; *or* built around a series of process and skill objectives related to small-scale problem solving exercises set firmly within the context of the everyday experience of the pupils. In either case the main emphasis would be on establishing a more structured understanding of the physical and biological world linked to an appreciation that scientific knowledge is potentially useful in day-to-day situations. As with the Phase 1 activities, a premium would be placed on encouraging pupils to be creative, speculative and able to relate their work in science to activities undertaken in other parts of the curriculum. In order to achieve these objectives we would recommend a substantial reduction in the content of courses; less concern for a sequential and incremental treatment of content; and the use of starting points for science studies that are directly within the first-hand experience of the pupils. While the laboratory will begin to be exploited as a valuable resource in the science teaching programme, we would hope that throughout this foundation course the initial stimulus for activities would be derived from the bio-physical and social environment of the school and its community. Thus actual phenomena, events and observations will act as "triggers" for hypothesis and investigation rather than relying on "experiments".

Years 3-5 *Optional Studies*. All pupils, irrespective of sex or differences in ability, would be required to study ONE and not more than TWO of the following courses [62]:

  (a) *Biological Science* ⎫ or *Integrated Science*
  (b) *Physical Sciences* ⎭
  (c) *Earth Sciences*
  (d) *History and Philosophy of Science*

Courses (a) and (b), which would largely be derived from the general ideas found in existing biology, chemistry and physics courses, would provide opportunities for young people to follow fairly traditional science studies to 16+ without the problems of choice and over-specialization that occur with the three-subject curriculum common in schools today. Course (c), Earth Sciences, based largely on biology, geology, meteorology and geophysics, would have to be developed in most schools and would serve as a broadening option for many youngsters. It would also act as a valuable link between the sciences and geography in the middle secondary curriculum. Course (d), based largely on the history and philosophy of scientific ideas, would provide a valuable bridge between science and the arts and social sciences, either as a course of study in its own right, or as a contextual course taken in conjunction with one of the empirically-oriented options (a), (b) or (c). As an alternative to the separate study of biological and physical sciences, a strong case can be made for the option of an Integrated Science course designed as either a one- or two-credit programme. A one-credit Integrated Science option would require very careful planning and development while the two-credit option could be easily developed from SCISP [63].

In terms of short-term curriculum development, options (a), (b) and (c) could easily be derived from existing syllabuses and reports provided a rigorous, and parsimonious, approach was adopted towards content selection and the definition of objectives. Option (d) would, on the other hand, require a major allocation of development time, given the low level of current concerns for the history and philosophy of science and the poor record of previous attempts to introduce this dimension into the science curriculum. At the same time we note the increased interest in this area in higher education during the last decade, and the interest aroused by recent attempts to develop more socially oriented science courses at the school level.

Phase 3

Years 6 and 7 *Optional Studies*. All pupils would be required to study ONE and not more than THREE of the following courses:
(a) *Biological Science*
(b) *Physical Sciences* or *Physics* or *Chemistry*
(c) *Earth Sciences* or *Geology*
(d) *History and Philosophy of Science*
(e) *Integrated Studies*
The range of options from the above that could be offered in any school would depend largely on the size of the sixth form, or its equivalent (e.g. a sixth-form college) and the nature of the examination system (e.g. an N and F system would allow much greater flexibility than the A-level system). Nevertheless we would support a further attempt to increase the acceptability of a Physical Sciences course and urge a simplification and reduction in the total content of advanced courses in Biology, Chemistry and Physics.

Summary

Model 1 is best summarized by the following table, which includes for Years 1–5 a recommended allocation of teaching periods (based on a 40-period week).

Phase 1 (5–11)	Introductory science studies integrated with the general curriculum	
Phase 2 (11–16)		
Years 11–13	Core Science (Foundation Course)	4 periods
	Optional Studies:	
Years 13–16	ONE or TWO from	
	Biological Science (4 periods)	
	Physical Science (4 periods)	
	Earth Science (4 periods)	
	History and Social Studies of Science (4 periods)	
	Integrated Science (4 periods)	4 or 8 periods
	or	
	Integrated Science (8 periods)	
Phase 3 (16–18)	*Optional Studies:*	
	ONE, TWO or THREE from	
	Biological Science (8 periods)	
	Physical Science (8 periods)	
	Physics (8 periods)	
	Chemistry (8 periods)	8, 16 or 24 periods
	Earth Science/Geology (8 periods)	
	History and Social Studies of Science (8 periods)	
	Integrated Science (8 periods)	

Comment
While a development along the lines of the above model would meet certain of the objectives we have considered earlier and might assist in broadening the approach to science studies, it does, nevertheless, suffer from certain inherent weaknesses. Although in theory there is no reason why the options (Years 13–16 and 16–18) should not be treated as being of equal value, in practice we doubt if this would be the case. It is in fact highly probable that the pattern of options would lead to able pupils undertaking certain options, while Arts-oriented students and, perhaps, the less able or poorly motivated students chose others. A more fundamental objection is that, even were this not to happen, the setting of one field of science studies against another in the way suggested implies that a broad general education in science can be achieved even when a key component such as, for example, biological science is not included. In considering this model it is important that the reader should bear in mind these limitations.

MODEL 2

Phase 1
Introductory Science as detailed on pp. 41–42.

Phase 2
Years 1 and 2 *Basic Science Core*. A two-year foundation course (as outlined in Model 1).

Years 3–5 *Core Science* (4 periods per week) plus ONE optional study drawn from either
(a) *Further Science* (4 periods)
(b) *Applied Science* (4 periods)
(c) *History and Philosophy of Science* (4 periods)

Core Science. The three-year compulsory Core Science would be concerned with two key issues, viz.
(a) an understanding of scientific knowledge, skills and processes; and
(b) an appreciation of the application of this knowledge in the everyday world.

In other words the core course would consist of an integrated study of elementary science and a broad introduction to science studies embodying the basic disciplines of biology, chemistry, geology and physics related throughout to a study of practical and technological applications and, to a more limited extent, historical and social implications. In designing core science programmes we recommend that a significant effort is made to explore the relationship between theory and practice by the careful utilization of case study techniques, field work, design projects and the use of practical problem-solving exercises. Pupils should be encouraged to develop both inductive and deductive strategies in their science studies, with an emphasis on developing a critical and creative analysis of the relationship between science and technology. Every opportunity should be taken to relate work in the field, the laboratory or the workshop to the world of work and leisure. In the final year of the core course we would urge the utilization of individual and small-group projects designed to explore in depth, and with rigour, some aspect of the relationship between science and technology.

Optional Courses. It is important to accept that each of the optional courses represents a tenable complementary field of study to the core course which will enrich and extend understanding of the relationship between science and technology. In general terms the distinctive characteristics of each option would be

(a) *Further Science*—a course designed to enable teachers and pupils to explore in greater depth the theoretical basis of some of the concepts of the core course. In this course a major emphasis would be placed on developing an understanding of the nature of scientific theory—its evolution, verification and utilization.

(b) *Applied Science*—a course designed to explore in greater depth the application of scientific ideas in technological and practical contexts. A major emphasis would be placed on design projects and the detailed study of science in use.

(c) *History and Philosophy of Science*—a course designed to explore the history of science and technology and the changing nature of scientific ideas and methodology. A thematic approach concentrating on theoretical and practical issues raised in the core course would be preferable to any systematic, and chronological, study of "the history of science".

Phase 3
Years 6 and 7   The sixth-form, or equivalent, course structure would be based on the three optional studies developed in Years 3–5, viz. *Further Science, Applied Science* and *History and Philosophy of Science*. We suggest that at this level a Major/Minor pattern is developed where a Major study involves 16 periods per week, and a Minor involves 8 periods per week. Students would be able to offer either

(a) a Major study (from Further Science, Applied Science, or History and Philosophy of Science);

(b) a Major plus a Minor; or

(c) one, two or three Minors. Students would not be able to study at both the Major and Minor levels in the same option. For curriculum design and examination purposes it might be convenient to define Further Science in specific discipline areas, e.g. Biology, Chemistry, Geology and Physics; or Biological, Earth and Physical Sciences (see Model 1).

Summary
The whole of Model 2 would involve a substantial reconsideration of content and teaching method and, at sixth-form level, a redefinition of examination objectives either in terms of A and S or N and F to accommodate the Major/Minor concept. We note however that this concept is common in the area of tertiary education, so the problems are not insuperable. Model 2 is summarized in the following table.

47

Phase 1 (5–11)	Introductory science studies integrated with the general curriculum		
Phase 2 (11–16)) Years 11–16 Years 13–16	Core Science (Foundation Course) Core Science: (4 Periods) plus one from		4 periods
	Further Science	(4 periods)	
	Applied Science	(4 periods)	8 periods
	History and Philosophy of Science	(4 periods)	
Phase 3 Years 16–18	Either 1 Major Study	(16 periods)	
	or 1 Major and 1 Minor Study	(8–24 periods)	
	or 1/2/3 Minor Studies	(8–24 periods)	8–24 periods
	drawn from	(24 periods)	
	Further Science		
	Applied Science		
	History and Philosophy of Science		

Comment

This model does represent a significant improvement on Model 1 in terms of the broad purposes of science education presented earlier. The introduction of a common core in Phase 2 overcomes some of the objections to the simple options systems but presents a new difficulty, namely that of developing a coherent core syllabus that effectively complements three diverse, and distinct, options. It can also be argued that a science curriculum consisting of core science plus further science will, in practice, be regarded as being of higher status than core science plus applied science. In short, the status and parity-of-esteem issues are not necessarily resolved in this model. Finally the model, as with Model 1, fails to achieve the broad coverage of aims and contexts because it contains an option system which enables essential aspects of a general science education to be excluded from the science curriculum for some, or in fact, all pupils.

MODEL 3

Our third, and final, model for the science curriculum is the simplest and yet the most radical in terms of current resources, expertise and assumptions. Starting from the same Phase 1 assumptions of experientially-based science studies, we suggest that Phases 2 and 3 be organized as follows:

Years 1 and 2 *Environmental Science* (8 periods per week)
Year 3 *Experiment Science* (8 periods per week)
Year 4 *Applied Science* (8 periods per week)
Years 5 *Science and Society* (8 periods per week)
Years 6 and 7 *Independent Studies* (16 periods per week)

Each of the above programmes would reflect issues discussed in the detailed analysis of Models 1 and 2, but this model represents the final move towards a common curriculum that meets the requirements of a "science education for all" previously described in terms of various combinations of core and option studies.

48

We are basically suggesting that, in the long term, the science curriculum should be conceptualized as five related **stages** that explore distinctive characteristics of scientific understanding. The model is incremental in terms of processes and contexts but not in terms of content. The essential characteristics of each stage would be as follows:

1. *Environmental Science*—an experience-based study of the immediate environment of the school designed to develop an understanding of basic scientific concepts and essential data-gathering and organizational skills.

2. *Experimental Science*—a period of extended laboratory work designed to develop a systematic approach and understanding of science as an empirical and experimental study. Themes, derived from Stage 1, would provide the starting point for this stage.

3. *Applied Science*—a period of extended laboratory and workshop activities intended to explore the application of scientific ideas generated in Stages 1 and 2 in the context of real-life problems and issues. This stage could be treated either thematically or in terms of specific design, development and evaluation projects.

4. *Science and Society*—the systematic analysis of the social implications of science and technology in the context of themes/projects developed in Stages 1, 2 and 3. At this stage the knowledge, concepts, processes and applications raised in the earlier parts of the course would be placed in a historical, social and personal context, i.e. the pupils would be required to evaluate the relevance to social issues of the scientific ideas they have developed, and to their own value systems.

5. *Independent Studies*—students, individually or in small groups, would select themes or topics for detailed investigation throughout the two years of the course. They would be required to integrate their investigations across the four basic elements identified in Years 1–5 and present substantial reports on their inquiries. An essential component of this stage would be the requirement that individuals or groups would report, at regular intervals, progress and problems to the wider audience of the total number of students involved in the course. They would, in short, be required to behave as scientists, i.e. expose, rationalize and justify their personal constructions of scientific reality [64].

While this model can be conceptualized as a series of distinct stages followed in sequence by all pupils, it could alternatively be treated as a range of perspectives differentially considered with respect to particular themes or concepts. Thus a concept such as "energy" could be progressively explored in environmental, experimental, applied and social terms during a specified period. This would then lead to a similar treatment of a related concept. In many respects this cyclic approach would be preferable to a more linear and compartmentalized presentation of the five stages. The cyclic approach would also enable concepts to be constantly modified and expanded in the light of additional knowledge and experience.

The implications of a movement towards a common science curriculum based on distinct activities related to specific contexts would have major implications in terms of curriculum design, development and evaluation; the initial and in-service training of teachers; the need to radically review and reconstruct the examination system, and the requirement to renegotiate the relationship between secondary and tertiary education. In suggesting that in the long term school science should move away from an incremental, and at times iterative, approach to content, and towards an overtly experiential and personal construction of scientific understanding, raises fundamental questions concerning the nature of educational processes, the purpose of educational institutions and the relationship between the learner and what is learnt. While we fully appreciate the immense difficulties of making real, even in a design sense, this third model we are convinced that no system of optional studies will in the end satisfy the broad requirements implicit in the notion of "science for all". By juxtaposing "pure" science, "applied" science and "socially applied" science in a range of options we immediately concede the possibility that

(a) one is more important than the other in terms of status, value or vocational goals;
(b) a general education, broadly conceived, can be complete without a concern for one or more of the options presented;
(c) pupils and students of differing initial abilities will be better served by a study of one option rather than another.

One alternative to the trap of the differentiated curriculum is the potential prison of a common curriculum defined in content terms—a curriculum devised in terms of lowest common factors and constructed in relation to what can be reasonably taught and learnt. Such a solution would, we suggest, be unacceptable to teachers, pupils, parents and employers. The only other alternative is that of placing a high premium on the creative ability of young people to make sense of their world and of ensuring that they have access to a scientific dimension related to commonsense knowledge and understanding. The potential freedom implicit in rejecting the dual yokes of linear mental development and the epistemology of school science is a daunting challenge, but one we feel science educators must at least consider during the next decade. The implications of this statement are considered in the next section of our report.

COMMENT AND CONCLUSION

We have described Model 3 as being the simplest and yet the most radical of our curriculum proposals. While in some senses this is true, it must also be noted that it is a "solution" to a curriculum problem that has been essentially defined in terms of current assumptions regarding schools, schooling, science and society. While the model may appear radical it has been described in terms familiar to all teachers and in a form that *could* be readily adopted in most schools. In evaluating the model as a solution to the future problems that may face science educators, it is perhaps important to reflect on changes in society that could occur during the next two decades and which would severely question the assumptions upon which Model 3 is based.

Model 3, as presented above, implicitly assumes that education is the prerogative of schools; that the young will always be educated in institutions called schools; that science education will always be concerned basically with supporting "big" science and "high" technology; and that the essential purpose of education is that of preparing people for a life at work. Even the most casual review of current social, economic and political thinking leads one to appreciate that each of these assumptions can be and is being questioned. One aspect of the current movement towards accountability in education is related to the issue of who should control education and to question whether it should be left in the hands of "educators". A substantial body of literature exists [65] which questions in the most fundamental sense whether schools, as institutions, are necessary or desirable. The possibility or inevitability of an energy crisis, and/or a nuclear holocaust, both question a blind allegiance to the pursuit of science and technology as currently defined in Western European terms. The next industrial revolution, based on the widespread application of microprocessors and the perhaps more esoteric aspects of genetic engineering, is likely to generate a redefinition of our common understanding of the relationships between work, leisure and life. It is not for us, in a document of this kind, or on the basis of our qualifications, to enter into any detailed discussion of the strengths, or otherwise, or the likely outcomes of the debates that surround each of the issues raised above. All we are qualified to do is to note possibilities and hazard a guess as to the implications for science teachers.

May we therefore conclude by stating that Model 3 in some form or other does provide a basis for inventive thinking about the future. Each of the five components would, and should, have some part to play in the education, or self-education, of the individual irrespective of social systems, institutions and the economic structure of any future society. Man will always need to understand and coexist with his environment; empirically test and evaluate his assumptions concerning his world; apply his knowledge to the processes of living; appreciate his cultural and historical predicament; and act creatively. He may, or may not, need to know or even rediscover Ohm's law. If these aspects of science education have at least in theory some relevance to the future as hypothesized, perhaps they should be seriously considered in the context of the here and now. To state our conclusions in these terms is not to appropriate for science education any special place in the curriculum of the future. We are merely acknowledging that it has, in common with other ways of looking at the world, an essential place.

IMPLICATIONS AND RECOMMENDATIONS

The earlier sections of this report have been concerned with a detailed critique of current provision in science education followed by a number of possible guidelines for future development. In attempting to draw together the trailed coats of the arguments and issues raised in the preamble to this final section we, as a small working party, have become increasingly conscious of the relative ease with which one can identify what is wrong with our current system of science education and the enormous problems of identifying ways of changing it—for the better. Even in the analysis of the current situation we have identified deep-rooted, and strongly argued, variations in interpretation of the current position; of the relationship between cause and effect; and the assumptions that can be made regarding the general trends in schooling, education and society. We have, however, agreed on one fundamental issue, namely, that science education in our schools cannot be considered in isolation from the curriculum at large and the tensions that exist between schooling, education and society. Arising from this basic agreement we wish to record our view

(a) that science education, particularly at the secondary school level and for the more able pupil, has improved considerably during the last two decades, and we should acknowledge our debt to all those who have contributed to the curriculum renewal movement;

(b) that many teachers have sought, often under difficult circumstances, to adapt existing resources designed for academically oriented pupils to meet the needs of the average and below average pupil;

(c) that the science teaching profession has in so many ways sought to adapt theory and practice to the changing context of comprehensive education for all, new examination systems and, in many respects, the hopes and aspirations of ordinary pupils, their parents and their reference groups.

At the same time we must also note

(d) that sciences syllabuses have become more heavily content laden, to the extent that at A level they contain much that previously would not have been encountered outside the confines of an honours degree course in science;

(e) that at the primary school level, which we regard as an essential building block in the total edifice of science education for all, we have, as a nation, failed to convert the hopes of the early 1970s into a reality for the 1980s;

(f) that school science has, both overtly and covertly, become more pure, conceptually demanding and complex, and less concerned with the everyday reality and experience of our youngsters, their parents and their employers: it has, in so many ways, become a complex symbolic system accessible to the few, and we do not believe that there is anything intrinsic in science, or science studies, that necessarily forces our subject in this direction;

(g) that while the needs of future scientists *may* be met within current provision, we have done little to relate science studies to other legitimate goals, particularly those related to education for life, for work, for citizenship or for leisure;

(h) that the important cultural aspects of science, its history, philosophy and contribution to the way twentieth-century man conceptualizes his environ-- ment, have not been adequately considered in the construction of examination syllabuses and courses at all levels of schooling;

(i) that science teaching, particularly at the secondary school level, too often

52

appears as a highly specialist and capital-intensive activity outside the general curriculum of the school. It appears, like modern languages, as a difficult but high-status adjunct to general education which, in the main, fails to create effective links with other subjects.

In seeking to make recommendations for the future we are convinced that there is a marked need for a broadly based debate on the future direction of science education and hope that this consultative document will at least provide a framework within which that debate can occur. In order to point up that debate we conclude by making the following assertions.

1. That science studies have a key role to play in general education but that radical changes in the content and nature of school science must be made if obligations are to be effectively discharged.

2. Young people, of all abilities and aspirations, have the right of access to the world of science, and it is incumbent upon all teachers and administrators to ensure that ways are found whereby science studies can become accessible to all.

3. That science education programmes and courses must be broadened at all levels to enable teachers and pupils to explore more flexibly and creatively the wider implications of science in society. This must mean a marked reduction in content and a more flexible and imaginative use of resources including the laboratory, the environment and the literature of science.

4. While experimental work and the detailed study of the conceptual processes that characterize science must remain central to all work undertaken, an equal, and balancing, emphasis must be placed on developing an understanding of the usefulness of scientific knowledge and processes in society and in everyday life.

5. As a direct consequence of the above assertion we recommend that substantial resources be allocated to a major programme of reseach and development that seeks to evaluate alternative definitions of school science; that develops and effectively evaluates curricula proposals in the areas of applied science, earth sciences and the history and philosophy of science; and which develops a series of small-scale and intensive studies of the nature of young people's conceptualizations of science and scientific processes.

6. That every opportunity is taken during the normal processes of curriculum review to move gradually in the directions suggested in this report. In particular we recommend that the introduction of a common examination system at 16+ should be seen as a major opportunity to realize some of the implications of our often repeated slogan, "Science education for all". Similarly we urge that irrespective of the outcome of the N and F debate, efforts will be made to improve A-level syllabuses.

Finally we urge the Association and all concerned with science education to consider in detail the implications of the HMI report *Primary Education in England* [66] and reconsider the basis upon which resources are allocated to this sector of educational provision.

7. That the Association, in cooperation with the LEAs, the Schools Council and the DES, responds as a matter of great urgency to the very considerable need to review science teacher training provision and the training of primary/middle school teachers at the initial and in-service levels. We feel that science teachers can respond to the repeated challenges referred to earlier only if realistic and practical provision is made for the further improvement of the teaching force, the resources available to it, and the conditions under which it works. While we do not wish to specify these matters in terms of teaching loads, class size, capitation allowances and study leave, we remain convinced that these factors must be constantly reviewed in the context of any proposal to change the curriculum.

8. Finally, we recommend that this consultative document is regarded by all who read it as a basis for discussion and not a prescriptive recipe for a brave, or frightening, new world.

As joint authors of this document we wish to support unanimously its publication, and unreservedly defend our individual right to enter the public debate we hope it will stimulate as potential sceptics of certain aspects of the analysis presented. The exercise has brought the working group together—the product and the issues we have discussed have rightly divided us.

NOTES AND REFERENCES

Part One   Science Education in Context

1. Science Masters' Association, *Science and Education: a Policy Statement* (Murray 1957).

2. *Higher Education:* Report of the Committee appointed by the Prime Minister under the Chairmanship of Lord Robbins, 1961–63 (London, HMSO 1963).

3. *The tables of O- and A-level passes in* Statistics in Education *(1952) show only Biology, Chemistry, Physics, and Physics with Chemistry at O level; and Physics, Chemistry, Botany, Zoology and Biology at A level.*

4. *1975 figures are quoted in Table 1 as these were the latest available at the time of writing. All statistics are from appropriate volumes of* Statistics in Education *(London, HMSO).*

5. Laybourn, K., and Bailey, C. H., *Teaching Science to the Ordinary Pupil* (University of London Press 1957).

6. *The UNESCO Source Book for Science Teachers* (UNESCO 1962).

7. *Typical of textbooks of this period are*
   (a) Pinsent, A. E., *The Principles of Teaching Method* (Harrap 1941).
   (b) Nunn, P., *Education: Its Data and First Principles,* 3rd edition (E. J. Arnold 1945).

8. *Children and their Primary Schools:* Report of the Central Advisory Council for Education under the Chairmanship of Lady Plowden (HMSO 1967).

9. *The Association has, in recent years, published a number of papers on science teaching in the middle years of schooling. These include*
   Science in the Middle Years (ASE Study Series No. 6 1976).
   Science in the Middle Schools: A Yorkshire Survey (ASE Study Series No. 5 1975).
   In addition, a number of the publications originating from the work of the Primary Schools Science Sub-Committee are relevant, e.g.
   Science and Primary Education Papers:
   The Present Situation: A Review (ASE 1976).
   The Role of the Headteacher (ASE 1976).
   A Post of Responsibility (ASE 1976).

10. Secondary Schools Examination Council, *The Certificate of Secondary Education: a Proposal for a new leaving certificate other than the GCE* (HMSO 1960).

11. *Of particular relevance to the discussion of science in the comprehensive school are the following Study Series papers:*
    Non-Streamed Science: A Teachers' Guide ASE Study Series No. 7 (ASE 1976).
    Non-Streamed Science: The Training of Teachers ASE Study Series No. 8 (ASE 1976).

Case Studies of Science Education in a Changing Context—an Interim Report by the Education (Research) Committee (ASE 1976, out of print).
The Lamp Project Publications (ASE 1976).
Non-Streamed Science Organization and Practice ASE Study Series No. 10 (ASE 1976).

12. *See, for example*
Collier, K. G. (Ed.), *Innovation in Higher Education* (NFER 1974); publications of the Society for Research in Higher Education, the Nuffield Higher Education Learning Project and Colleges of Education Learning Project.
In addition to innovative approaches to teaching and learning in further and higher education, it is important to note the recent introduction of a wide range of multi- and inter-disciplinary first degrees in British polytechnics and universities.

13. *The pattern and nature of higher education in the 1980s and 1900s will largely depend on the planning decisions taken as a result of the Government's discussion paper*
Higher Education into the 1990s: a discussion document (DES and SED 1978).
Of the five planning models described in this document only Model E is likely to lead to a major change in the recruitment pattern for higher education in the general direction of increasing the proportion of students from the lower socio-economic groups.

14. *Education: A Framework for Expansion* (HMSO 1972)

15. *For an interesting discussion of this issue see*
Hall, S., "Education and the Crisis of the Urban School" in *Why Urban Education?* Ed. J. Raynor (Open University 1974).

16. (a) Braverman, H., *Labour and Monopoly Capital* (Monthly Rev. Press 1974).
(b) Whitty, G., and Young, M. F. D., *Explanations in the Politics of School Knowledge* (Nafferton Books 1976).

17. *Perhaps the most significant indicator of the movement towards greater accountability is the recent establishment of the Assessment of Performance Unit by the Department of Education and Science. The initial thinking of the Science Working Group is outlined in*
Assessment of Scientific Development (Assessment of Performance Unit 1977).

18. *See for example*
(a) Hull, C. L., *Essentials of Behaviour* (Yale University Press 1951).
(b) Skinner, B. F., *Cumulative Record* (Methuen 1962).
(c) Thorndike, E. L., *Human Learning* (Cornell University Press 1931).

19. *See*
van Praagh, G. (Ed.), *H. E. Armstrong and Science Education* (John Murray 1973).

20. *The work of Dewey and Piaget is widely known and acknowledged as a major factor in changing attitudes and practice in primary and early secondary education.*

Less well known is the contribution made by the English writers Nathan and Susan Isaacs. The interested reader is referred to

Isaacs, S., *Intellectual Growth in Young Children* (Routledge 1930).

Isaacs, S., *The Children We Teach* (University of London Press 1932).

Isaacs, N., *Early Scientific Trends in Children* (National Froebel Foundation 1960).

21. *Both* Science 5/13 *and* SCISP *are based on psychological models of human development. The former project relies on an interpretation of the work of Piaget on the intellectual development of children, while the latter is linked to Gagné's concept of learning hierarchies; see*

Gagné, R. M., *The Conditions of Learning* (Holt, Rinehart and Winston 1970).

22. *The reader interested in recent commentaries and criticisms of Piagetian psychology is referred to*

(a) Jenkins, E. W., "Piaget and school chemistry—a critique", *Education in Chemistry*, May 1978.

(b) Brown, C., and Desforges, C., "Piagetian psychology and education: time for revision", *British Journal of Educational Psychology*, 47, **7**, 1977.

23. *This is a complex area to discuss within the limitations of a short discussion paper on the future directions science education might take. The interested reader should, therefore, explore a few of the texts listed below and note that Mead and Kelly represent respectively social–psychological and psychological models of human development that nevertheless have as a common focus the development of individual views of personal reality. These relativistic models contrast sharply with the deterministic models that have dominated educational thinking to date. It is also important to note that the gradual rejection of a reductionist view of science itself equates in an interesting way with the gradual emergence of reflexive and humanistic models of personal development and intellectual growth. See*

Mead, G. H., *Mind, Self and Society* (Chicago, University of Chicago Press 1934; paperback edition 1962).

Mead, G. H., *The Philosophy of the Act* (as above).

Kelly, G., *The Psychology of Personal Constructs,* Vols. 1 and 2 (New York, Norton 1955).

Bannister, D., and Fransella, F., *Inquiring Man* (Penguin 1971).

Salmon, P., "Education in the light of personal construct theory", *Education for Teaching*, Summer 1974, pp. 25–28.

24. *Apart from the Schools Council Integrated Science Project (SCISP), the only attempt to devise a science curriculum on the basis of process objectives linked to other curriculum areas is the recently published*

N and F study in Integrated Science, *Schools Council 18+ Research Programme: Science*, 1977.

Other developments such as the JMB A-level syllabus in Environmental Science and the Malvern/ASE Science in Society Project, basically fail to come to terms with the problem of integrating science studies with work in "subject" areas such as history, geography, literature, art and aesthetics; and the "value" areas of politics and social theory. In general terms most attempts to liberalize the science curriculum have foundered on the rock of technological application before they have encountered the iceberg of social context (i.e. the interrelationship between science and the society that generates and supports it). From a wide range of available literature on the interaction of science and society we would particularly note:

Barnes, B., *Scientific Knowledge and Sociological Theory* (Routledge and Kegan Paul 1974).

Bernal, J. D., *The Social Function of Science* (Routledge and Kegan Paul 1939).

Bernal, J. D., *Science in History* (Penguin 1969).

Ravetz, J. R., *Scientific Knowledge and its Social Problems* (Penguin 1971).

Rose, H., and Rose, S., *Science and Society* (Penguin 1970).

Richardson, M. N. (Ed.). *What is Science?* an introduction to aspects of the philosophy and sociology of science (ASE, in preparation).

25. *The major exceptions to this are the Nuffield Junior Science Project; the Schools Council Integrated Science Project, and at the sixth-form level the Nuffield Working with Science Project.*

26. *Appropriate references to the philosophical analysis of the nature of knowledge are*
    (a) Peters, R. S., *Ethics and Education* (George Allen and Unwin 1966).
    (b) Hirst, P. H., and Peters, R. S., *The Logic of Education* (Routledge and Kegan Paul 1970).
    (c) Hirst, P. H., *Knowledge and the Curriculum* (Routledge and Kegan Paul 1974).
    (d) Phenix, P. H., *Realms of Meaning* (New York, McGraw-Hill 1964).

27. *Bernstein's related concepts of "classification" and "framing" are well described in his contribution to*
    Young, M. F. D. (Ed.), *Knowledge and Control* (Collier-Macmillan 1971).
    The argument in this paragraph relates to philosophical position of the writers such as Hirst and Peters to the sociological analysis of Bernstein in the general context of the position science occupies in the secondary school. The boundaries between "science" and "non-science" are drawn both in terms of curriculum content and the identifiable resources of school science, viz. laboratories, preparation rooms, technicians, storerooms, etc. These physical resources reinforce the intellectual separatism of science from the "common culture". It is, perhaps, significant that the same separatism is found in institutions of higher education including the new universities. The binary divide is perhaps the ultimate manifestation of the gap between the "pure" and "applied" concepts of education.

28. Schools Council Bulletin No. 3, *Changes in School Science Teaching* (Evans/Methuen Education 1970), p. 7.

29. *The basic dispute between Karl Popper and Thomas Kuhn on the nature of scientific knowledge is to be found in the following texts*
    (a) Popper, K., *Conjectures and Refutations* (Routledge and Kegan Paul 1963).

(b) Kuhn, T., *The Structure of Scientific Revolutions,* 2nd edition (University of Chicago Press 1970).

(c) Lakatos, I., and Musgrave, A., *Criticism and the Growth of Knowledge* (Oxford University Press 1970).

For a commentary and analysis of these issues as they affect the science teacher, please see

(d) Richardson, M. N. (see Note 24).

(e) Squires, A., *Science in the Middle Years* (ASE 1976).

Part Two   The Current Curriculum

30. Science Masters' Association, *Science and Education: a Policy Statement* (Murray 1957).

31. *Higher Education:* Report of the Committee appointed by the Prime Minister under the Chairmanship of Lord Robbins, 1961–63 (London, HMSO 1963).

32. See Note 10.

33. *See for example*

(a) Booth, N., "The Impact of Science Teaching Projects on Secondary Education", *Education in Science,* June 1975.

(b) Nicodemus, R. B., "Discrepancies in Measuring Adoption of New Curriculum Projects", *Education in Science,* November 1975.

(c) Schools Council, *Impact and Take-up Project* (University of Sussex. Primary school survey completed 1978, secondary survey 1979).

34. *The major example of the Association working through local teacher groups has been the LAMP Project directed by R. Driver and J. Nellist. In addition to a series of materials this project has published two teacher guides*
   A Teachers' Handbook: A Guide to the Project (ASE 1976).
   Teachers' Handbook 2: The Organization of Modular Courses (ASE 1978).

35. See *The Head of Science and the Task of Management* (ASE 1978).

36. The ASE *Science and Primary Education* papers listed in Note 9.

37. See Note 35.

38. *Education in Schools: A Consultative Document* (HMSO 1977).

39. *Curriculum 11–16.* Working papers by HM Inspectorate: a contribution to current debate (HMSO 1977).

40. *See for example*
   Non-streamed Science: A Teachers' Guide (ASE 1976).
   Non-streamed Science: The Training of Teachers (ASE 1976).

41. *Resource Based Learning: A Teachers' Guide* (ASE 1978).

42. *A Language for Life:* Report of the Committee of Inquiry under the Chairmanship of Sir Allan Bullock (HMSO 1975).

43. *Language in Science Education,* ASE Study Series (ASE, in preparation).

44. Evetts, Sir J., Address to the Royal Aeronautical Society Conference, Cheltenham, 1952.

45. *Schools and Industry* (CBI 1977).
46. Working Mathematics Group.
47. *Project Technology Handbook* (Heinemann 1972).
48. *Teachers in Industry Project* (CBI 1977).
49. The Science Lessons from Industrial Processes Scheme in Sunderland.
50. See *Closer Links between Teachers and Industry and Commerce*: a joint report by the CBI and the Schools Council in association with the Scottish Education Department (HMSO 1966).

Part Three Curriculum Proposals

51. Kaplan, A., *The Conduct of Enquiry* (Chandler, Intertext 1964).
52. *We have referred to these aims as being person centred to avoid any confusion with the concept of student-centred aims or objectives, which often suggests that the curriculum is organized around student interests or teachers' perceptions of student "needs". Related concepts to a person-centred approach are the development of self-awareness and the self-concept.*
53. Bullock, A., "Science—a tarnished image?", *School Science Review*, 201, **57**, June 1976.
54. *It is important to emphasize this point. The working party responsible for this paper had neither the expertise nor the resources to produce outline syllabuses or teaching materials to support our notions of curriculum models. At this stage we can only suggest possible lines for debate and development as any attempt to be prescriptive, or even overtly illustrative, would, in our view, be premature.*
55. *This factor is, in our view, qualitatively different from the consideration of the relationship between scientific knowledge and society alluded to in paragraph (c). In paragraph (f) we are concerned with the fundamental nature of all knowledge, including scientific knowledge, and the debate as to whether knowledge can be considered to exist outside the consciousness of man. Interested readers are referred to*
    (a) Schutz, A., *The Phenomenology of the Social World* (Heinemann 1972).
    (b) Berger, P. L., and Luckmann, T., *The Social Construction of Reality* (Penguin Books 1966).
56. *Science for the Under-Thirteens* (ASE 1974).
57. *We would argue strongly that all primary and middle schools should be provided with a wide range of pupil books related to science studies as well as Teachers' Guides and other materials developed by such projects as*
    (a) *Nuffield Junior Science (Collins 1967).*
    (b) *Science 5/13* (Macdonald 1972–75).
    (c) *Project Environment* (Longman 1974).
    (d) *Environmental Studies 5/13* (Hart-Davis 1972).
58. *Science for the Under-Thirteens*, op. cit., pp. 24–25.

59. Bruner, J. S., *Towards a Theory of Instruction* (Harvard University Press 1966).

60. *We are fully aware that Bruner's proposition is derived from the field of developmental and theoretical psychology and is, in the view of many, highly controversial. It is, however, a concisely worded alternative proposition to those that suggest that different forms of knowledge are accessible only to young people of specific abilities or social backgrounds; or to pupils who have reached a particular stage of psychological development. For contrasting views the reader is referred to*
   (a) Bantock, G. H., *Culture, Industrialisation and Education* (Routledge and Kegan Paul 1968).
   (b) Jensen, A. R., *Educability and Group Differences* (Methuen 1973).
   (c) Hutchinson, M., and Young, C., *Educating the Intelligent* (Penguin Books 1962).

61. *We are here referring to the wide range of general, or combined, science courses that have been developed to overcome the requirement to introduce the specialist study of biology, chemistry and physics at age 11. Among these we might highlight*
   (a) Mee, Boyd and Ritchie, *Science for the Seventies* (Heinemann 1971).
   (b) *Nuffield Combined Science* (Longman 1970).

62. *The introduction of the term "courses" is quite deliberate for we are not referring to "subjects" called biological science, physical science, etc. Each course would utilize appropriate concepts, skills and techniques derived from subjects such as biology, chemistry, physics, geology, meteorology, history, philosophy, sociology, etc.*

63. *We fully acknowledge the practical difficulties of discriminating between a one- and two-credit Integrated Science course and refer to SCISP an an illustration of the problem.*

64. *We note that several current A-level courses contain opportunities for extended project work and this proposal is a simple, and logical, extension of such developments.*

65. *See for example*
   Holt, J., *How Children Fail* (Penguin 1969).
   Holt, J., *Instead of Education* (Penguin 1977).
   Illich, I., *Deschooling Society* (Penguin 1970).
   Reimer, E., *School is Dead* (New York, Doubleday 1971).

66. *Primary Education in England* (HMSO 1978).

SOME PUBLICATIONS OF THE ASSOCIATION FOR SCIENCE EDUCATION

Study Series

		£
No. 1	The Place of Science in Environmental Education	0.30
No. 5	Science in the Middle Years: A Yorkshire Survey	0.55
No. 6	Science in Middle Schools	0.65
No. 7	Non-streamed Science: A Teacher's Guide	1.00
No. 8	Non-streamed Science: The Training of Teachers	0.80
No. 9	The Supply of Science Teachers	0.40
No. 10	Non-streamed Science: Organisation and Practice	1.20
No. 12	Developments in Science Teacher Training	0.60
No. 13	The Head of Science and the Task of Management	1.30

Science and Primary Education Papers

No. 1	The Present Situation: A Review	0.10
No. 2	The Role of the Headteacher	0.35
No. 3	A Post of Responsibility	0.50

All the books in these series are printed and published by the Association for Science Education and are available from ASE, College Lane, Hatfield, Hertfordshire, AL10 9AA.

LAMP Project

	£
Teachers' Handbook 1	0.75
Teachers' Handbook 2	0.80
Topic Briefs:	
No. 1 Fuels	0.80
No. 2 Heating and Lighting a Home	0.80
No. 3 Pollution	0.70
No. 4 Materials	1.20
No. 5 Photography	0.60
No. 6 Gardening	0.70
No. 7 Health and Hygiene	1.75
No. 8 Space and Space Travel	1.60
No. 9 Paints and Dyes	0.80
No. 10 Flight	0.80
No. 11 Science and Food	0.70
No. 12 Science and the Motor Car	0.80
No. 13 Problem Solving	0.50
No. 14 Fibres and Fabrics	N.Y.P.
No. 15 Electronics	N.Y.P.

All the books in this series are printed and published by the Association for Science Education and are available from ASE, College Lane, Hatfield, Hertfordshire, AL10 9AA.